Contents

1980s

1990s

2000s

Acknowledgements

50 Best Business Ideas is certainly a collaborative effort. Conceived in 2010 by David Lester, the book has only made it through to publication because of the significant input of many people.

At the outset, David's initial brainstorm provided a skeleton list of ideas. This was embellished by business school professors, investors and successful entrepreneurs, as well as members of Crimson's business editorial team. I would also like to extend particular gratitude to Kamal Ahmed, business editor at the *Sunday Telegraph* for his valuable input.

Sara Rizk followed the initial brief to perfection to create a template chapter on the modest Post-it note and, alongside me, commissioned some of the early chapters. Others at Crimson, namely Stephanie Welstead, Gareth Platt and Georgina-Kate Adams, worked tirelessly in the latter stages to elevate the stories from a potted history to something of meaning and usefulness for lovers of business and great ideas.

Many journalists contributed chapters, namely Gareth Platt (8), John O'Hanlon (8), Ryan Platt (8), Jon Card (6), Emma Haslett (3), Georgina-Kate Adams (2), Henrietta Walsh (2), Hannah Prevett (2), Nicole Farrell (2), Hugh Jordan (2), Sara Rizk, Peter Crush, Carys Matthews, Trevor Clawson, Mark Shaw and Martin James.

Picture research for *50 Best Business Ideas* was co-ordinated by Beth Bishop and Lucy Elizabeth Smith, with much of the time-consuming and determined research carried out by Lucy, Abigail Van-West and Francesca Jaconelli. Between them, they unearthed a treasure trove of incredible images dating back to the very first products that made it to market, and their enthusiasm and effort is worthy of thanks. Equally, Samantha Harper painstakingly fact-checked each and every chapter.

I would also like to thank a number of others at Crimson who have helped to ensure this book made it through the final stages, namely Lucy Smith, Gemma Garner, Clare Blanchfield and especially Jonathan Young in marketing, and Jo Jacomb and Dawn Wilkinson in production. Dawn's patience, in particular, is worthy of extra appreciation here. Knowledgeable colleagues such as Hugh Brune, Trudi Knight and Kevin Paul carried out some crucial sense-checking of initial chapter drafts. And finally, I would like to thank my wife and three sons, who were incredibly supportive throughout.

Introduction

For thousands of years, ingenious minds have grappled with the world's conundrums, desperate for the solutions capable of making a tangible difference to the way we live and work.

50 Best Business Ideas celebrates the greatest business ideas of the past 50 years. In the past half-century alone, some highly influential ideas and innovations have emerged that have had a profound impact on business and society. Why the last 50 years? As a period, it rivals the industrial revolution for economic advancement. The 'computer age' has enabled instant access to information and knowledge. The processes of miniaturisation and digitisation have made advanced technology more portable. And globalisation and commoditisation have dramatically enhanced the commercial potential of any market innovation.

Each of the ideas has been undeniably transformative. Many remain core to our daily existence: the PC, the internet, satellite TV and barcodes, to name a few. Some are controversial: the disposable nappy, the contraceptive pill, aerosol deodorant and plastic bags have had an enormous impact, irrespective of the negative press they have received. And there are ideas that have declined in influence already: such as the fax machine, the Post-it note, the Walkman and the pocket calculator, which were outmoded by digitised versions, but nevertheless critical as evolutionary landmarks.

In identifying and acclaiming the individuals and companies that changed the world, we offer a nod to the incredible foresight of the people who had the germ of an idea and set the ball rolling.

Great stories

There are some fantastic stories contained within this book, some of which you may have heard of, but also many that will surprise and enthral, if only for the way such major contributions to the world came about through happenstance. For example, the inventions of Kevlar, smoke alarms, microwave ovens and the Post-it note, as we know them now, had an element of serendipity about them, only happening because scientific experiments went wrong. Kevlar, the polymer used in body armour and innumerable other applications for its impervious qualities, was stumbled upon when scientist Stephanie Kwolek of DuPont decided not to throw away the cloudy, runny substance and instead spun it into a fibre to see what happened. It turned out this fibre was stronger than any other previously created – and one now responsible for a multi-billion dollar income for licensor DuPont each year.

Likewise, the smoke alarm was created when a detector for poisonous gases failed to produce the desired response. When Swiss physicist Walter Jaeger decided to take a cigarette break from his toils, he noticed the smoke had triggered the sensors. While it wasn't he who went on to create the domestic smoke alarm we know today, the timeframe for a commercially viable version would surely have been more prolonged without his discovery. And the Post-it note was ultimately arrived at after Dr Spencer Silver's adhesive failed to bond fully with other surfaces. When colleague Art Fry used the weak glue to stick small pieces of paper as temporary bookmarks in his hymn book, the Post-it note was effectively born.

The idea for a ring pull on food and drink cans emerged from the mind of Ermal Fraze when he forgot his beer can opener while having a picnic in 1959. It took him a few years and some effort, but when the 'pull-tab' launched (or 'snap top' as it was known then) it was an instant success, and has had a fundamental impact on the consumption of food and drink the world over. Another food-related invention, the microwave oven, only transpired because of a partially melted chocolate and peanut bar. Dr Percy Spencer of defence contractor Raytheon stood next to a magnetron system during testing in 1944 and found that his snack bar had inadvertently been heated by the experiment. The company patented the idea for the use of microwaves in the cooking process in 1945 and invented an unwieldy oven some years later. It wasn't until 1965 that a worthy product was created for domestic use, however.

Regarding the invention of fibre optics, it's telling that the first presentation suggesting light could be used as a transmission medium was laughed at by professional engineers. It didn't stop (now Sir) Charles Kao from pursuing such an unlikely dream. Showing similar perseverance, VHS only came to market when it did because two employees at JVC carried on working on a de-prioritised project. And most of us have heard about (now Sir) James Dyson's 5,000+ prototypes prior to the launch of his Dual Cyclone vacuum cleaner.

For those who enjoy tales of conflict, there's the story of the three scientific teams who went to war over who had the right to proclaim themselves the inventor of Magnetic Resonance Imaging (MRI), and there's the German-Brazilian Andreas Pavel who fought Sony for 20 years for the right to be known as the inventor of a Walkman-like device, a battle the Japanese conglomerate settled out of court. Pavel ran up a $3.6m legal bill and came close to bankruptcy for the recognition he felt he deserved. You'll undoubtedly find intrigue, inspiration and considerable amounts of information in each of the stories.

Methodology

Our final 50 is without parallel. We employed influential minds from the fields of academia, business and finance, who provided valuable input and discussed the project at length. We have identified truly seminal moments rather than pre-cursors that had limited commercial impact. So, while early forms of the fax machine were operational in the 1920s and attempts were made to create a smoke alarm in the 1890s, we decided to include them because these were not the break-through moments, the catalysts or game-changers in their own right.

In most cases, examples like this didn't reach far beyond the garden shed or laboratory. The feverish minds that contemplated their creation's impact on the world so often succumbed to time's relentless march and an inability to deliver something the rest of the planet appreciated. So, before every good home or office could benefit from their wild imaginations, the baton was passed to others, culminating through iterative innovations in what we see and use today.

There are subjective choices for which we offer no apology. We know the Pill celebrated its 50th birthday last year, but crucially it was not marketed as a contraceptive until a year hence. There are others, too, that some might argue pre-existed our cut-off. But look for the evidence that society had embraced the concepts, that money was flowing into the coffers of the trailblazers, and you'll find very little. They don't pass the commercial viability test that we applied and outline in each chapter.

As with the vacuum cleaner, it's not always the inventor that warrants all the credit; it's often the innovators that come later and make a material difference to the early concept. By tweaking, deconstructing and reconstructing, by testing and tampering, these people and companies revolutionised things that were, at best, satisfactory. That is why this book recognises specific ground-breakers, such as the Walkman, the Dyson and Google's '20% innovation time', alongside ideas exploited simultaneously by a multitude of leading players.

Conversely, we acknowledge key moments that took place years, even decades, before the product's commercial explosion. The internet is a good example of this: although millions now credit Sir Tim Berners-Lee as the 'father of the web' for his scientific work in the late 1980s and early 1990s, the decisive moment happened much earlier when ARPANET, a computer network created for the use of the Pentagon in the US during the 1960s, was first deployed. Without Berners-Lee's World Wide Web and his gift of hypertext mark-up language (HTML), ARPANET would not have made billions for the founders of Google and Facebook or the pocket change users of eBay enjoy; but his invention required the backbone of ARPANET to have true significance.

There are lists within lists, such as ideas that arguably made the world more secure (smoke alarms, Kevlar, biometrics and remote keyless entry systems); more fun (the MP3 player, the computer game, satellite TV and VHS); more efficient (the barcode, the spreadsheet, the fax machine, video conferencing, email, fibre optics, computer-aided design (CAD), the PC, pocket calculators, Just-in-Time inventory management, electronic point of sale technology, smartphones, tablet computers, pay-by-swipe technology, search engines, GPS and ATMs); more consumer-friendly (the ring pull, e-readers, the Dyson, touch-tone phones, infrared remote controls, digital cameras, the compact disc and microwave ovens); made business unusual (the 20-70-10 rule, Google's 20% innovation time, and pay-per-click advertising); and more equal (Equal Opportunities policy and budget airlines).

Inevitably, there are some that ostensibly don't fall into any of the above categories, such as the hybrid car, and many that you could argue fall into multiple pots. And there were also those that missed the cut; ones that would have laid strong claim to being as important as any we actually selected. The microchip (also known as the integrated circuit) is only just the wrong side of 50, as is the laser (the first functioning laser was operated in 1960, much to our chagrin).

Ultimately, it's a unique and fascinating read. Contained within these pages are the definitive stories of the ideas that shaped the last half-century. The pre-cursors, personalities, patents, prototypes, prolonged legal battles, and lucrative proceeds of the blood, sweat and tears that have gone into each and every idea, are all in here.

This book puts its head above the parapet and so our 50 is designed to be debated long and hard, and we welcome such discourse. Enjoy.

1960s

1
The Pill

When: 1961

Where: USA

Why: The invention of the contraceptive pill gave women around the world control of their sexuality and was a major step forward in family planning

How: Much prior research into hormones was aimed at improving fertility, but the combination of oestrogen and progestin produced an effective contraceptive

Who: Carl Djerassi, Frank Colton

Fact: In the UK, a quarter of 16- to 49-year-old women take contraceptive pills, which come in 32 different forms

Since its launch half a century ago, the contraceptive pill has been credited with everything from triggering sexual anarchy and torpedoing family values to transforming the complexion of the workplace, in allowing women to enter board-level corporate positions for the first time. Some believe it increases a woman's risk of contracting cancer; others believe it provides a crucial defence against this disease.

Everyone, it seems, has a different take on the Pill's social impact. What is not in doubt, however, is its importance as a commercial idea. More people have taken the Pill than any other prescribed medicine on the planet; today, more than 100 million women use it regularly. In 2010 the contraceptive pill market was worth more than $5bn, and this figure could rise further still as the morning-after pill gains traction in the developing world.

Many experts believe that, for all the revolutionary advances made by the Pill so far, its full commercial and social potential has yet to be realised. In 2010, one influential female commentator said that the provision of oral contraceptives in the developing world could save the lives of 150,000 women, and 640,000 newborns. With researchers making excellent progress on a male contraceptive pill, it's safe to assume the market for this product will change dramatically in the years ahead.

The background

It was in the 1930s that hormone researchers first observed that androgen, oestrogens and progesterone inhibited ovulation in women. The development of a practical medication based on hormone therapy was delayed because of the high cost of obtaining these chemicals, so one of the principal goals was to find a way of synthesising them so that they could be manufactured in bulk.

During World War II, Russell Marker, an organic chemist at Pennsylvania State University, founded Syntex in Mexico City to exploit his discovery that Mexican yams could be used to synthesise progesterone. Syntex became the focus of the commercial development of the essential raw materials for pharmaceutical contraception, attracting three of the most significant figures in the history of the Pill.

The first of these was George Rosenkrantz, who took over from Marker in 1945 and retraced Marker's process steps. He was aided by Carl Djerassi, a chemist, novelist and playwright who joined Syntex in 1949, and the Mexican chemistry professor Luis Miramontes, who joined the following year. Though Djerassi is sometimes called the father of the contraceptive pill, it took a team effort along with Rosenkrantz and Miramontes to synthesise the first orally highly active progestin norethindrone in 1951.

During World War II, Russell Marker, an organic chemist at Pennsylvania State University, founded Syntex ... to exploit his discovery that Mexican yams could be used to synthesise progesterone.

That was 10 years before the launch of Enovid by D. G. Searle. Frank Colton had joined Searle as a senior research chemist in 1951; the following year he synthesised the progestin norethynodrel, an isomer of norethindrone, which combined with the oestrogen mestranol. The resulting drug, Enovid, was approved by the US Food and Drug Administration (FDA) in 1956 for menstrual disorders, and gained further endorsement in 1960 as the world's first oral contraceptive following trials initiated in Puerto Rico by Dr Gregory Pincus and supervised by Dr Edris Rice-Wray in 1956. It is clearly difficult to decide whether Colton or the Syntex team should take the credit for inventing the

Marcia Goldstein, the publicity director of 'Planned Parenthood', examines their birth control marketing material.

contraceptive pill, but Colton's efforts earned him induction into the USA's National Inventors Hall of Fame in 1988.

Marketing of the Pill started in the USA and the UK in 1961. The Combined Oral Contraceptive, as well as many other methods of birth control, was legalised in Canada in 1969. The market was ready to embrace this liberating new technology. In the 1960s, women liked the Pill because it provided a reversible method of birth control that was, and still is today, almost 100% effective when taken as directed.

Commercial impact

After gaining approval from the FDA in 1960, the Pill quickly gained traction in the American market. By 1964, 6.5 million American women were on the Pill; and this figure had almost doubled by 1968. Although the authorities initially tried to put a brake on the spread of oral contraceptives by permitting access to married women only, this just resulted in creating a thriving black market. Far from being deterred, young, single women came to see the Pill as alluringly edgy, and the oral contraceptive soon became an essential accessory for women caught up in the hedonism of the mid-1960s. Meanwhile, in Britain, the story was similar: the number of British women using the Pill rose from 50,000 in 1962 to over 1 million in 1969, as the Beatles-led revolution of youth took hold.

By 1964, 6.5 million American women were on the Pill; and this figure had almost doubled by 1968 ... the number of British women using the Pill rose from 50,000 in 1962 to over 1 million in 1969.

Other countries were more reluctant to embrace the oral contraceptive. In Japan, for example, the country's Medical Association prevented the Pill from being approved for nearly 40 years, citing health concerns. The world's Catholic population has remained reticent on religious grounds – this is perhaps most clearly evident in Italy, where just 20% of women aged 15 to 55 use any kind of contraception. However, despite these pockets of resistance, the morning-after contraception market has continued to expand steadily. Today the Pill provides the cornerstone of the global contraceptives market, which, according to analysts, is growing at an annual rate of 4%.

The growth of the birth-control pill has certainly changed the shape of the contraceptive industry, with several alternative products suffering a drop in market share as a result. The diaphragm was an early victim of this trend: in 1940, one in three American couples had used a diaphragm for contraception; while in 1965, this figure had fallen to just 10%, and continued to plummet thereafter. By 2002, it was just 0.2%. The move towards oral contraception has also had severe repercussions for the condom industry. Although in Britain condoms remain the most popular form of contraception because of their protection against sexually transmitted infections, sales have been gradually falling worldwide: in India, for example, condom usage has fallen dramatically in recent years, largely because the country's young people believe the Pill offers greater convenience and a better sexual experience.

A far more wide-reaching consequence of the Pill has been the enhancement of female opportunities in education and business. It is widely accepted that oral contraception has given women more control over when they have children, allowing them to plan high-powered careers before marriage and motherhood. According to a 2002 study led by Harvard University, the number of female graduates entering professional programmes increased substantially just after 1970, and the age at which women first married also began to rise around the same year. Progress has remained steady since then; in fact, the proportion of women on the boards of Europe's top companies increased from 8% in 2004 to 12% in 2010. If the current rate of growth is maintained, gender parity could be achieved by 2026.

A far more wide-reaching consequence of the Pill has been the enhancement of female opportunities in education and business ... According to a 2002 study ... the number of female graduates entering professional programmes increased substantially just after 1970.

It is difficult to put a financial value on the impact of women in the workplace. However, no one could deny that women have invigorated the world of business with talent, enthusiasm and a fresh perspective. In 2006, *The Economist* claimed that women had contributed more to global GDP growth over the preceding 20 years than either new technology or the new economic super-powers, China and India. Given that some of the world's top companies, such as Yahoo!, Pepsi Co, Kraft Foods and DuPont, now have female CEOs, it's easy to see why women have had such an impact.

What happened next?

The landscape of the contraceptive pill market is currently changing dramatically. As patents begin to lapse, the three major players – Bayer, Johnson & Johnson and Warner Chilcott – are having their dominance eroded by the emergence of new, generic contraceptives. Bayer continues to lead the market, having achieved sales in excess of $1.5bn last year. However Bayer's patents in both Europe and America have recently expired and the company has been involved in an expensive monopolisation lawsuit in the US – a clear sign that its dominance is under threat.

With over-population becoming increasingly problematic in Third World countries, use of the contraceptive pill seems set to increase further.

Furthermore, the market is continuing to expand, with more and more women in the developing world turning to the Pill to avert unwanted pregnancies. In 1960, fewer than 10% of unmarried women in the developing world used contraception; this number had increased to 60% by 2000, and the Pill was firmly entrenched as the most popular contraceptive method. With over-population becoming increasingly problematic in Third World countries, use of the contraceptive pill seems set to increase further.

This increase is sure to bring controversy. Indeed it has been suggested that the world's deadliest killer, HIV/AIDS, has spread faster as a direct result of oral contraceptive programmes in those countries where it is endemic. The Population Research Institute says that, to date, more than 50 medical studies have investigated the association of hormonal contraceptive use and HIV/AIDS infection. These studies show that hormonal contraceptives increase almost all known risk factors for HIV: upping a woman's risk of infection, increasing the replication of the HIV virus, and speeding up the debilitating and deadly progression of the disease. A medical trial published in the journal *AIDS* in 2009, which monitored HIV progression according to need for antiretroviral drugs (ART), stated: 'The risk of becoming eligible for ART was almost 70% higher in women taking the (contraceptive) pills. '

Furthermore, the link between contraceptive pills and abortion remains a source of contention. Perceived wisdom holds that the rise in use of the Pill has led to a fall in abortion rates; however, many contemporary studies dispute this. A study carried out in Spain, published in January 2011, found that the use of contraceptive methods increased by over 30% from 1997–2007, and that the elective abortion rate more than doubled over the same period. This throws fresh doubt over the Pill's ability to control procreation and avert unwanted pregnancy.

However, it seems that these controversies will do little to diminish the Pill's popularity. Researchers are thought to be on the brink of unveiling a new male pill, which will work by inhibiting sperm output, to the point of temporary infertility. Indeed, in July 2011, researchers at Columbia University in the USA claimed they had found a solution that mimics vitamin A deficiency, a common cause of sterility. The development of the male contraceptive has attracted widespread concern, with many experts raising the possibility of negative reactions to alcohol; but it seems that a product will arrive on the market in the not-too-distant future.

2
The disposable nappy

When: 1961

Where: UK

Why: An idea of pure convenience, which millions of parents around the world choose not to do without

How: A Procter & Gamble chemical engineer picked up on an unpatented design and turned it into one of the world's most successful industries

Who: Vic Mills

Fact: Approximately eight million disposable nappies are used in the UK every day

There have undoubtedly been more glamorous innovations than the disposable nappy, but few can have had more of an impact on everyday household life. At a stroke, the throwaway nappy, or diaper to use its American name, solved the age-old problem of how to keep young children clean and hygienic, without the need for constant washing and replenishing of their dressings – which had taxed mankind since the dawn of time.

Like many great ideas, the disposable nappy was laughed out of town when it first appeared on the scene. However, a sharp-eyed manager at Procter & Gamble saw its benefits, and used it as the foundation for one of the 20th century's most enduring and commercially successful business ideas.

The disposable nappy is not without detractors – environmental groups claim that at least four-and-a-half trees are required to keep a single baby in disposable nappies before they are potty trained, while critics also point out that it costs the British taxpayer £40m to dispose of all Britain's used nappies each year. However, research has also shown that due to the level of energy required to wash cloth nappies, the CO_2 equivalents are actually higher for cloth nappies than disposables. In addition, as cloth nappies are less absorbent and more likely to cause discomfort and nappy rash, according to scientific research, they have to be changed more often. Even some committed users of cloth nappies switch to disposables at night.

Few can deny that the throwaway nappy has made life for new mums more convenient, less time-consuming and far more hygienic than before.

The background

Parents have used protective clothing for children who aren't toilet trained since time immemorial – but it wasn't a pretty business. Our ancestors used everything from leaves to animal skins to swaddle their babies, while the word 'diaper' derives from a material used to clothe infants during the Middle Ages. In the late 19th century, Americans began to use cloth nappies, while British people used a similar garment known as a terry nappy; these garments could be reused and washed several times, and were held in place with a safety pin.

Parents have used protective clothing for children who aren't toilet trained since time immemorial – but it wasn't a pretty business.

A major breakthrough in disposable nappy technology came just after World War II, when an American housewife called Marion Donovan invented the

'Boater,' a diaper with snaps for fastening and a waterproof cover. In 1949, her waterproof nappy was launched by Saks Fifth Avenue in New York, and had instant commercial success.

But Donovan had less success in persuading retailers to buy into her other big idea – a disposable nappy. None of the major firms in New York saw the value, or feasibility, of a disposable nappy, so Donovan decided to concentrate on her waterproof version instead – eventually selling the patent to children's clothing specialist Keko in 1961.

However, from the late 1950s another visionary was already working on taking forward the disposable diaper. In 1957 Vic Mills, a legendary chemical engineer with Procter & Gamble, who had already transformed the production process for ivory soap and paved the way for Pringles crisps, was tasked with creating new product lines for the company's recently acquired paper business in Green Bay, Wisconsin. Drawing on his own experience as a grandfather, Mills decided to refine and adapt Donovan's disposable dream.

Using a doll that allowed water into its mouth and secreted it at the other end, Mills and his team set to work. Pushing the doll on a treadmill, the researchers tested various materials to create the perfect diaper. By 1961, they had their design: a simple rectangular nappy made of rayon, tissue and plastic, held together with safety pins. All that was needed was a name. Then Alfred Goldman, creative director at ad agency Benton and Bowles, hit upon 'Pampers'. It was swiftly adopted.

Procter & Gamble introduced Pampers, the first affordable and successful disposable nappy.

Commercial impact

Demand for the new disposable nappies increased steadily as people cottoned on to the benefits they offered. Soon competitors began to enter the market, notably Huggies, which was brought to market by Kimberly-Clark, a leading consumer brands company.

The commercial impact was clear – disposable diaper products soon became Procter & Gamble's single biggest brand, with annual sales growth reaching 25%. Before long, the industry was worth $3bn a year.

The rate of innovation in the disposable nappy market was matched by the speed of its market growth.

Procter & Gamble now had to battle with Kimberly-Clark, Johnson & Johnson and a host of other companies to maintain its market share. Meanwhile, behind the scenes the early diaper design was being modified to produce a lighter, simpler and more user-friendly product.

Among the key innovations that followed was the introduction of a 'third size' by Pampers in 1969. A year later, tape to fasten the nappy was added to the design. Then, during the 1980s, super-absorbent polymers were developed, which reduced the nappy's weight by around 50%.

The rate of innovation in the disposable nappy market was matched by the speed of its market growth. In 2008, Global Industry Analysts estimated that worldwide sales of disposable diapers would reach $26.6bn by 2012 – and $29bn by 2015, with further growth being driven by the rapid population growth in the Asia-Pacific region.

Britain remains a substantial consumer of disposable nappies, and last year it was estimated that Pampers' share of the UK market was worth £480m. In 2004, Pampers became the first Procter & Gamble brand to achieve turnover in excess of $5bn and in 2010 it was identified as the UK's most trusted brand, in a survey produced by Millward Brown and the Futures Company.

What happened next?

The invention of Pampers was the last great achievement of Vic Mills' legendary engineering career, and he eventually settled into retirement in Arizona, where he lived to the age of 100. As for Marion Donovan, she went back to university and earned a degree in architecture from Vale, Colorado in 1958. She went on to earn more than a dozen patents.

Looking forward, the market position of the disposable nappy looks set to endure. However, as ethical issues become a wider topic of conversation, disposable nappies do face competition from some more earth-friendly competitors. Reusable cloth nappies, which can be washed after use, have come back into favour.

Furthermore, many regions offer nappy-laundering services, which take used cloth nappies away each week for cleaning and replace them with fresh ones. This service can now be offered for roughly the same price as a week's supply of disposable nappies. However, for most parents, the convenience of disposable nappies remains a priority, and thus biodegradable nappies seem to be the best-placed innovation to take the disposable nappy into the 21st century.

3

Contact lenses

When: 1961

Where: Prague, Czech Republic

Why: Soft contact lenses have brought comfortable and flexible vision to millions of people around the world

How: A prototype spin-casting machine, made using a child's building set, and parts from a bicycle, provided the mechanism for the manufacture of the first soft contact lenses ever sold

Who: Czech chemist Professor Otto Wichterle

Fact: 125 million people use contact lenses worldwide

Every morning, 125 million people around the world stick their fingers into their eyes. While it may make you cringe, for these contact lens wearers the benefits of improved comfort and flexibility are significant enough for them to choose this option over wearing spectacles.

There are many reasons why people wear lenses – for some it is a matter of appearance, while many favour their practicality, particularly when engaging in sporting activities. Furthermore, unlike spectacles, they are not normally affected by wet weather, don't steam up and provide a wider field of vision.

Although contact lenses have been manufactured since the late 1800s, it was not until the introduction of soft contact lenses that the market really escalated. The launch of disposable lenses in 1988 created a renewable commodity, which could generate a sustainable income for retailers – marking their position as a significant business innovation.

The background

The concept of the contact lens can be traced back as far as Leonardo Da Vinci. In 1508 Da Vinci detailed a method of altering corneal power by submerging the eye in a bowl of water, in 'Codex of the Eye, Manual D'. However, it was not until the late 1950s that a significant breakthrough in the conception of soft contact lenses was made. Professor Otto Wichterle, a Czech polymer chemist, and his assistant, Dr Drahoslav Lím, developed a transparent hydrogel plastic named 'hydroxyethylmethacrylate'. This was composed largely of water and was soft and pliable when wet, but was capable of drying to become hard.

Although Wichterle was a director of the Czechoslavak Academy of Sciences (CSAS), at that time the Institute of Macromolecular Chemistry was still being built, so the earliest experiments were carried out at his home. The scientists published their research, 'Hydrophilic gels for biological use', in the journal Nature in 1960, but the design was not ready to market yet.

While Wichterle's soft and permeable fibre was comfortable to wear, its high water content made it difficult to handle. Furthermore, it provided a poor optical quality and debates arose about its potential to absorb infectious bacteria. Wichterle continued his research and in 1961 constructed a prototype of a spin-casting machine, using his son's building set and parts from a bicycle. This centrifugal casting procedure provided a new way of manufacturing lenses, enabling the product to enter the market at last.

However, without Wichterle's knowledge, CSAS sold the patent rights to the US National Patent Development Corporation. Thus, it was in America that the spin-cast 'Soflens' material first obtained approval, from the US Food and Drug Administration (FDA). Bausch & Lomb made the first soft, hydrogel lenses commercially available in the USA in 1971. A year later, British optometrist Rishi Agarwal proposed the idea of disposable soft contact lenses, although this product was not launched on the UK market until 1988.

Wichterle continued his research and in 1961 constructed a prototype of a spin-casting machine ... This centrifugal casting procedure provided a new way of manufacturing lenses.

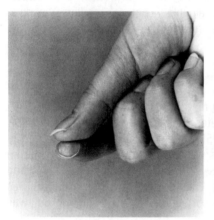 

The 1971 Hydrogel Contact Lens.

Commercial impact

The introduction of soft lenses to optics triggered a significant boom in contact lenses across the board. Their increased comfort and versatility saw them swiftly overtake their hard predecessor, and today only 2% of the contact lenses sold in the USA are hard lenses (although 16% of the US market uses gas-permeable hard lenses – a 1979 innovation that allows oxygen to reach the cornea, just as a soft lens does). In 2010 the US soft lens market was estimated to be worth $2.1bn (£1.3bn), whilst the overall contact lens market had a worldwide value of $6.1bn (£3.7bn). According to Global Industry Analysts Inc, if the market continues to grow as predicted, this figure could nearly double by 2015; giving the global contact lens market a value of $11.7bn (£7.2bn). Leading companies in the market include Johnson & Johnson, which Baird & Co estimates controls more than 40% of the market, with Ciba Vision, CooperVision, and Bausch + Lomb among other key players.

The number of businesses selling contact lenses and providing optometry services has soared, as more retailers have sought to gain a share of this expanding market. Guernsey-based opticians and contact lens retailers Specsavers joined the sector relatively late, in 1984. However, the company grew rapidly, through a franchising system, to generate a 2008–09 income of £1.02bn. Today it has 1,390 outlets globally, including stores in the UK (one out of every three UK contact lens wearers chooses Specsavers), the Channel Islands, Ireland, the Netherlands, Scandinavia, Spain, Australia and New Zealand and employs over 26,500 people. But that is just the tip of the iceberg. The World Council of Optometry acknowledges 250,000 optometrists in 49 countries around the world – a number so high that they are represented globally by 96 membership organisations.

Opticians are best placed to profit from the contact lens market because they can pitch the product to customers while reviewing their prescription and assist with any related issues. Thus, many began retailing their own-brand lenses. However, some contact lens manufacturers, such as Acuvue, benefited from stocking their products in optometrists' surgeries.

But it was not just retailers who benefited from the boom in soft contact lenses. As demand grew, so too did the level of investment in lens innovation. The polymers from which soft lenses are manufactured continued to improve, as manufacturers altered the ingredients to enhance oxygen permeability. In 1980 the first tinted lenses came to market. They were coloured light blue to make them easier for wearers to see, in the event that they dropped them.

The innovations that followed came thick and fast. In 1981 seven-day extended wear contact lenses were introduced. These could be worn continuously for seven days without having to be removed at night. The following year, bifocal lenses were launched. However, it was in 1987 that the most significant innovation occurred. The manufacture of the first disposable contact lens created a new market in which contact lenses were a renewable commodity, providing retailers with an excellent opportunity to gain a sustainable income through repeat sales and customer loyalty. Some opticians, such as Specsavers, signed customers up to direct debits under which they made quarterly deliveries of disposable lenses to their customers' doors.

The main difference between regular contact lenses and disposables is that, while the former need to be removed (generally every night) to be cleaned and soaked in saline solution, their successors can be thrown away after use. 'Daily Disposables', which were launched in 1995, now have 57% of the UK market, while 2001 saw the emergence of Extended Wear Soft Lenses, which could be worn continuously for 30 days before being removed and disposed of.

The manufacture of the first disposable contact lens created a new market in which contact lenses were a renewable commodity, providing retailers with an excellent opportunity to gain a sustainable income through repeat sales and customer loyalty.

However, in the 1990s and 2000s the disposable soft contact lens sector became riddled with controversy after it was claimed that daily and extended wear lenses were the same product sold in different packaging. The cost of long-life contact lenses, such as 7- or 30-day extended wear products, could be as much as five times higher per day than that of daily lenses, leaving many customers feeling that they'd been scammed.

Several high-profile court cases ensued and Bausch & Lomb was forced to pay $68m in cash and vouchers to 1.5 million customers. Likewise, in 2001, Johnson & Johnson (which owns Acuvue) faced a legal battle in the USA and subsequently offered substantial compensation to customers who paid more for long-life lenses in the mistaken belief that the two products carried different medical recommendations.

In 2005, the UK newspaper the *Daily Mail* claimed that British contact lens wearers could be paying an additional £250m a year as a result of the scam. However, disposable soft contact lenses remained popular, revealing that in the 21st century consumers see contact lenses not as a luxury purchase, but as a necessity.

What happened next?

The popularity of soft contact lenses continues to be strong and the market has in fact expanded beyond those who have genuine eyesight difficulties. The introduction of non-prescription coloured lenses has seen the item become a fashion accessory, allowing consumers to change their eye colour just as they would dye their hair. However, there are concerns about the effect on overall eye health of wearing cosmetic lenses, as they can easily be purchased online without consulting an optician. This is part of a wider market of online-only contact lens retailers, where businesses battle to provide the lowest price, taking a share of the income that walk-in opticians would otherwise have received.

However, the greatest threat to the soft contact lens industry is the availability of laser eye surgery. This has the potential to eliminate the market for lenses, and indeed glasses, by correcting poor eyesight, thus making such items redundant. Yet this does not seem likely in the short term, as there is still much debate about the safety of the procedure and the long-term results it delivers. Furthermore, the high cost of the surgery prices many consumers out of the market. By contrast, contact lenses continue to provide a flexible product at a price that is affordable for the masses.

4

Satellite television

When: 1962

Where: USA and Europe

Why: Satellite TV allows widespread delivery of programming without the need for expensive infrastructure

How: Rapid advances in satellite technology meant that science fiction became reality

Who: NASA and private consortiums

Fact: By far the biggest satellite TV provider in Europe, Sky TV is in 36% of UK households

Walk down any street in the UK, and if you didn't spot at least a few of the distinctive dishes mounted to the sides of houses, it's fair to say something would be amiss. With the potential to bring a complete viewing experience to even the most obscure corners of the globe, and enabling instantaneous live transmissions in a way once thought impossible, the satellite has changed the way we think about television.

The background

The first person to suggest the idea of satellites being used for the relay of information was in fact the science fiction writer Arthur C. Clarke in 1945, 12 years before the first satellite was even sent into orbit. In his piece entitled 'Extraterrestrial Relays', he suggested that information could be beamed up to space stations in orbit and dispersed over a wide area, enabling instantaneous worldwide communication. The article was purely speculative – but uncannily similar to how satellite television ended up operating.

The first-ever orbital satellite, Sputnik I, was launched by the Soviets in 1957, followed closely by the USA's Explorer 1 in 1958. Almost immediately, scientists began exploring the concept of using these heavenly bodies to relay data over a wide area, using the satellites as a 'mirror' to reflect focused rays sent from Earth and dispersing them across the globe.

Just a few years later, in 1962, concept became reality, with the first-ever satellite television signal broadcast from Europe across the Atlantic via the AT&T-owned Telstar satellite. At the same time, NASA was experimenting with

JPL team (part of NASA) works on complex pieces of Explorer 1.

the concept of 'geosynchronous' satellites – satellites that would move at the same speed as the Earth's rotation, enabling them to 'hover' over the planet and deliver a constant, uninterrupted communication to a specific area. The first such satellite, the Syncom 2, was launched a year later, in 1963.

Hot on the heels of these early experiments came the next rapid advances towards satellite television – the launch of the first commercial communications satellite (IntelSat 1) in 1965, followed by the first national satellite-based television network in 1967 (the Soviet-run Orbita) and the first commercial television satellite launched from North America (Canada's Anik 1, launched in 1972).

The early satellite broadcasters relied on the only model that was feasible at the time – they would broadcast direct to Earth-based receiver stations owned by the cable companies, who would then deliver the picture via the older cable system to their subscribers. The first of these satellite-to-cable transmissions came in the landmark year of 1976, when Home Box Office (HBO) broadcast the heavyweight boxing match between Frazier and Ali, 'The Thrilla in Manila', to subscribers all over the USA.

The first of these satellite-to-cable transmissions came in the landmark year of 1976, when Home Box Office (HBO) broadcast the heavyweight boxing match 'The Thrilla in Manila' to subscribers all over the USA.

The year 1976 was a landmark year for another reason; it marked the inception of Direct-to-Home broadcasting (DTH), a fundamentally different way to deliver television to consumers, and still the basis for the current satellite television model. It came from an unlikely place – the garage of Stanford professor and former NASA scientist H. Taylor Howard. Whilst experimenting with video transmissions, Howard created for himself a large receiver dish made out of used microwave parts. He found that the dish was able to pick up the satellite signal intended for the cable companies and used his new invention to watch the cable service HBO for free. He attempted to reimburse them, sending them

The first geosynchronous Satellite.

a cheque for $100 to cover the movies he had watched – only to be returned the cheque and informed that HBO was concerned only with the large cable companies and was not seeking remuneration.

This innocuous rebuttal proved to be a huge missed opportunity for television networks, as the next years saw the explosion of free satellite television, a model that left them firmly out in the cold. H. Taylor Howard co-founded Chapparal Communications in 1980 with engineer Bob Taggart in order to sell their home satellite receiver dishes. When they began, the dishes they sold were prohibitively expensive at over $10,000 – but six years later the average price of a dish had dropped to around $3,000, the free satellite television market was booming and Chapparal was worth $50m. For customers, the advantages over cable television were plain; the initial set-up costs may have been much higher, but in return the customer was able to receive over 100 channels from different providers, apparently for ever, at no cost – and the removal of the intermediary cable service also meant significantly better picture quality.

All of this, of course, horrified the networks, who were incensed that the service they were providing at high cost to themselves was being watched by millions of people, none of whom provided them with a single penny. They set about trying to reclaim these lost profits by lobbying governments, and in 1984 were granted permission to encrypt their satellite feeds through the US Cable Act. Starting with HBO in 1986, this was the beginning of the end of satellite dish owners being able to watch television for free. This, along with an accompanying crackdown on illegal decoders, saw the market for satellite systems plummet during the next four years, as cable made a resurgence.

By 1990, however, with free satellite television well and truly killed off, the stage was set for companies to provide their own brand of licensed direct-to-consumer satellite television – the Direct-Broadcast Satellite (DBS). DBS used more powerful satellites, meaning that much smaller dishes were needed; from just 18 inches in diameter as opposed to the two- to five-metre giants needed to receive Taylor's DTH. The mini-dish was connected to a set-top box which decoded the feeds, usually through a card plugged into the device that was normally paid for with subscription charges. Starting with Primestar's inaugural service in the USA, this marked the moment when satellite television became a major commercial success for the providers.

Commercial impact

The history of direct-to-consumer satellite television in the USA is told through several companies: initially Primestar pioneered the service, but it gave way in the mid-1990s to DirecTV and others.

In the UK, however, commercial satellite television has been dominated since its birth by one name: British Sky Broadcasting (BSkyB), the corporation that continues to be the market leader in the sector by some distance. Sky Television Plc, owned by News Corporation's Rupert Murdoch, which owns Fox in the USA, started its DBS service in 1989, broadcasting four free channels from the Astra satellite, but the service was not popular and failed to return a profit. A year later, in an attempt to shore up its finances, Sky decided to join forces with fellow struggler British Satellite Broadcasting. The merger was a success and very likely rescued the fortunes of both companies; whilst Sky's overheads were much lower, because it did not own or maintain its satellite and was based in an industrial estate, it was let down by a lack of advertising revenue, something British Satellite was able to provide with its more prestigious contacts. BSB's own satellites were shut down, the service was moved to Astra, and BSkyB was born.

In 1992, Sky secured a major coup by gaining exclusive live broadcast rights to the inaugural FA Premier League in a £340m deal.

The next few years saw changes and a rapid phase of growth for BSkyB, admittedly helped along by hefty injections of News Corporation money. In 1991, the Sky Movies channel became the first to use encryption technology to operate along a subscription model, with others soon following. In 1992, Sky secured a major coup by gaining exclusive live broadcast rights to the inaugural FA Premier League in a £340m deal; this represented a significant bargaining chip (Murdoch described it as a 'battering ram') to attract subscribers. The year 1992 also marked a turning-point for the company as it began to turn in an operating profit. An indication of how much Sky had grown in just six years came in 1996, when over half a million subscribers tuned into the Bruno vs Tyson heavyweight boxing match, broadcast on pay-per-view.

A further milestone was 1998, when the Astra 2A satellite was launched in a new orbital position. This allowed Sky to begin transforming its previously analogue service to Sky Digital, a new way of broadcasting that had the potential to deliver many hundreds of video and audio channels to consumers; and paving the way for high-definition broadcasts in the future. A year later, it increased sales with a free mini-dish-and-digibox deal, and by 2003 it passed the remarkable milestone of seven million subscribers.

With the company firmly established as top dog, it began to focus on greater innovation in its products – Sky+, launched in 2001, allowed subscribers to pause, rewind and record live TV, a feature that now ships as standard on many pay-TV packages, and in 2004 it was the first in the UK to introduce broadcasting in high definition. In 2010, it was also the first to launch a 3D service.

What happened next?

British Sky Broadcasting is still growing, and continues to lead the market. In 2010 it passed 10 million subscribers, the first and currently only company in Europe to do so, and is now in over a third of households in the UK. In June 2010 it bought out rival Virgin Media Broadcasting for an estimated £160m, further consolidating its advantage over its competitors. In that same year, it reported revenue of almost £6bn; of which £855m was profit.

In fact, Sky's dominance has been such that concern over the health of the industry has arisen, with a 2010 Ofcom report requiring the company to wholesale its exclusive sports and film channels to other providers such as ITV and Virgin, after it found that there was a danger of a monopoly developing. Further issues have arisen over majority owner Rupert Murdoch's bid to mount a full takeover of the company, taking it off the stock market; numerous media groups have lobbied the government over the issue, describing the potential sale as a serious threat to media plurality.

And it's not just Sky's power that is a source of concern for the satellite TV industry. Over half of households have satellite in the UK, and there are worries that demand may be reaching a plateau, with the growth in subscribers slowing down across Europe. After countries complete their digital switchover, people will be able to watch a greater selection of channels than was offered by terrestrial satellite TV and may feel less of a need to switch to satellite.

The traditional subscription payment model of satellite TV may also see a change in the coming years, with the return of the early 1980s model in Freesat, a service provided by BBC and ITV, which allows customers to receive satellite channels for free after paying for the dish and receiver (Sky also has a similar service). Although any sustained threat to paid satellite from Freesat is in its infancy, with just over 1% of households using it, it may be the case that the value offered by such services will eventually undermine the paid subscription model.

A real threat to satellite TV comes with the rise of the internet. Previously, of course, customers were restricted to watching television on their TV set and simply had a choice between terrestrial, cable and satellite. But with both BT and Virgin rolling out high-speed fibre optic lines across the country, television delivered via the web is becoming a reality; Sky already allows all its subscribers to watch a selection of channels live online, and the major networks now operate on-demand online services where viewers can revisit shows at their leisure. It could be that in the future, given that it is already in many homes, the web may prove to be the smart choice for watching television for many; diminishing satellite's influence.

The traditional subscription payment model of satellite TV may also see a change in the coming years, with the return of the early 1980s model in Freesat ... which allows customers to receive satellite channels for free after paying for the dish and receiver.

As the world is introduced to high-speed internet, it also marks the beginning of a new challenge to satellite in fibre optic television. The technology takes advantage of the bandwidth offered to deliver greater interactivity than can be offered by satellite, giving the customer more control. Currently, Virgin is the only operator of fibre optic, and again the technology is just beginning to be used, so it remains to be seen whether this is a serious game-changer; but the concept of delivering internet, phone and TV through one simple port, without the need for installation of a dish, may well turn out to be an appealing one for many people.

In the face of all this, one thing seems clear – BSkyB's adaptability and continued innovation are set to continue its strong presence into the future. As for satellite TV itself, it remains the most popular choice for television, but only time will tell whether it will weather the storms of new technologies in the years to come.

5

Biometrics

When: 1962-63

Where: Laboratories and research institutes around the world

Why: The use of exact body matching offers an unrivalled level of security for businesses of all sizes

How: Driven by the needs of law enforcement agencies, researchers have worked on a range of methods to identify and authenticate individuals through unique characteristics

Who: Various

Fact: Biometrics is widely used in security and has potential for internet and mobile commerce dentification and authentication lie at the bedrock of the modern economy. When you log on to a computer network or slide your debit card into a cash point the software begins by asking two crucial questions. Who are you? And are you really who you say you are? Biometrics – the science of identifying individuals through unique characteristics – provides an automated means to answer those questions.

The background

The most common tools to establish and authenticate identity – notably passwords, PINs and swipe cards – are all highly fallible. Passwords are routinely shared or written down on Post-it notes for all to see, while PIN numbers and swipe cards can be stolen easily. For much of the 20th century, a passport or driving licence was required when you wished to pay for goods without cash; the former option was extremely inconvenient and left the carrier vulnerable to theft, while the latter was impossible for those who didn't drive.

Biometric technology, in theory, provides a much more accurate way to identify and authenticate. All biometric technologies are based on the single principle that we can establish the identity of individuals by means of measuring designated physical (or even behavioural) characteristics and checking those measurements against a database. It's certainly not a new concept. The invention

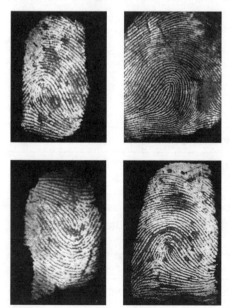

of fingerprinting in the 1890s was one of the first modern examples of systematic biometric identification. In more recent times, the science of biometrics has blossomed.

Today there are systems designed to recognise people by fingerprints, patterns in the iris, blood vessels in the retina, facial characteristics, voice characteristics and DNA. The technology has become increasingly widely used since the 1960s; the catalyst has been the rapid increase in computer processing power and storage capacity that we've seen over the last four decades. This computational power means that technology such as facial recognition and iris scans has become a reality.

Fingerprints made visible with ink.

The invention of fingerprinting in the 1890s was one of the first modern examples of systematic biometric identification.

The earliest examples of biometric technology were driven by the demands of law enforcement. In the 1800s, French anthropologist Alphonse Bertillon developed a system of body measurement that was used by police forces to identify suspects. The system fell out of favour when it became apparent that these measurements were by no means unique to each individual. Enter Scotland Yard's Richard Edward Henry, who came up with the much more effective system of fingerprinting. For the next 30 or 40 years, fingerprinting was pretty much the only game in town, but then, in 1936, ophthalmic technician Frank Burch proposed the use of iris patterns as an alternative means to establish and authenticate identity.

Fast-forward to the 1960s, and the search for biometric solutions was intensifying. In 1962, Woodrow W. Bledsoe developed a semi-automatic face-recognition system for the US government. A year later, Hughes Research Laboratories published a paper on fingerprint automation, an idea that the FBI was championing by the end of the decade.

The years that followed saw the introduction of a range of new technologies, with landmarks including the award of US patents for iris, speech and handprint recognition systems in the 1980s. By the following decade, biometric technology had emerged from the laboratory and was beginning to make an impact in the real world. In one of the most high-profile examples, a hand geometry system was introduced to control access to the Olympic Village in Munich in 1996. Another sporting-event application was at the Tampa Super Bowl 2001, police used a face-recognition system to identify criminals trying to enter the stadium.

In 2003, the US government-backed National Science and Technology Council established a sub-committee to plan and coordinate research and development and collaboration at national and international levels. The European Union also established a biometrics forum, with the aim of making the continent a world leader.

And it was soon clear that biometrics would impact on all our lives. For instance, in Britain, the compulsory identity cards planned by the Labour government were to contain a biometric chip to prevent identity theft. In 2005, the Home Office embarked on a roadshow that set out to explain why the use of biometrics was a good thing. The reasons included identity protection, fraud reduction and border security.

Meanwhile, the US government said that it would require biometric information on the passports of foreign visitors seeking to use the visa-waiver system when entering the USA. The result was that all passports subsequently issued in Britain contained a simple biometric chip. And while the national identity scheme was abandoned, millions of Britons become part of the biometric revolution every time they pick up their passports and head for the airport.

Commercial impact

As biometric technology becomes cheaper, it seems that the market is set to go from strength to strength. In fact a report from Research and Markets forecast that the biometric industry would grow at a rate of 22% per year until 2014. Biometrics firms are particularly buoyant in Japan; a 2008 report claimed that the industry was worth around $60m, with some technologies growing by up to 200%.

The rapid development of biometric technology has affected business in a number ways. First and foremost, the growing interest of governments in the science is providing a huge amount of funding for universities and small private companies – such as Digital Signal Corporation, a 3D face-recognition company based in the USA, which completed a $15m investment round in 2011.

Much of the demand for biometrics has thus far come from governments, specifically their security and law enforcement arms – for example, it is reported that the USA's Pentagon has set aside $3.5bn for biometric technology between 2007 and 2016.

But the potential of biometric systems goes far beyond the security and law enforcement concerns of governments. Seamless and accurate authentication and identification has many useful applications for private sector businesses.

US Marine Corps Sgt. A.C. Wilson uses a retina scanner to positively identify a member of the Baghdaddi city council prior to a meeting with local tribal figureheads, sheiks, community leaders and US service members deployed with Regimental Combat Team-7 in Baghdaddi, Iraq, on 10 January, 2007. Wilson is attached to the 4th Civil Affairs Group.

With USB fingerprint scanners now available for as little as £50, just about anyone can protect the data on their PC through this once-expensive technology.

As in the public sector, security is a key driver. For most of us the username/password system is the key that both allows us into computer networks and defines the information that we are permitted to view. But passwords are only as secure as the people using them choose to be. Research by Microsoft suggests that the average British worker has to retain around 20 password and username configurations for personal and business use, so it's hardly surprising that a significant number of us write the details down and often leave them in full public view. A much simpler solution is a switch to biometric systems such as fingerprint recognition. It's easier for the individual – not so much to remember – and far more secure for the organisation.

Biometrics is particularly attractive to businesses operating in sectors where the data is critical and sensitive. Examples include banking giant HSBC, which is now rolling out a facial recognition system to identify staff and contractors in its data centres. In Japan, 60,000 ATMs have been fitted with vein biometric technology – which uses the veins in a person's body to verify their identity.

Biometric protection is not just for large organisations. With USB fingerprint scanners now available for as little as £50, just about anyone can protect the data on their PC through this once-expensive technology. Meanwhile, in the mobile sphere, facial recognition apps are available as an alternative to passwords.

The applications for biometrics extend beyond security. Indeed, we're already seeing examples of biometric systems underpinning e-commerce. The Parent Pay system is a case in point. Used by schools in the UK, Parent Pay allows parents to create accounts for their children which can be loaded with money by debit or credit card. Once the account is loaded, children can pay for items such as school meals by placing their fingers on fingerprint readers.

Once the identity of the pupil is established, anything that is bought during the session is debited from the live account. There are a number of advantages to the system, notably that neither the school nor the pupil has to handle money on a daily basis. If the scheme were based around keying in passwords or PINs, there would probably be a lot more resistance from pupils who could well find it easier to hand over coins. As it is, it provides a convenient way of buying food.

What happened next?

As the technology beds in we're certain to see more variations on the e-commerce theme. For example, Nick Ogden, the entrepreneur who set up the Worldpay system – which allows even very small-scale retailers to take payment via credit and debit cards – has recently launched Voice Pay. Just as it says on the tin, the technology will identify individuals by means of their voice patterns, with the authentication process facilitating commerce via mobile phones.

But biometric technology isn't perfect. Changes in hairstyle, facial hair, weight and the use of make-up can fool face recognition systems, and tests have shown that fingerprint scanners can be 'hacked' by fake fingers. Even the patterns on the iris can be changed by some kind of traumatic accident. As a result, mission-critical biometrics may require multi-test systems involving more than one identifier. At present that often means the combination of one biometric and one non-biometric item to reduce the risk of fraud. For instance, passports (non-biometric) now have biometric chips included.

As the technology moves towards the holy grail of 100% accuracy, we are likely to see more applications for biometrics, particularly in the area of e-commerce. For the moment, though, governments and large organisations are the most enthusiastic users.

Originally conceived in 1979 by British entrepreneur Michael Aldrich, e-commerce systems built upon the innovations of EDI and Electronic Funds Transfer (EFT) systems to allow consumers to interact directly with businesses using automated systems. Seen in a limited form in the 1980s with ATM cash machines and telephone banking, e-commerce took off in the mid-1990s with the introduction of high-speed internet protocol and the entry of pioneering online businesses such as eBay and Amazon.

After the dotcom crash of 2001, in which many e-commerce businesses failed because of unrealistic business models, many retailers with a strong offline presence moved into the field, and e-commerce is now big business – in 2010, online shoppers bought over £590bn worth of goods.

EDI in its original, leased-line form is being slowly phased out in favour of internet-based systems, but the revolution it brought about in business is here to stay, and its influence will be felt for a long time to come.

In 1991, the Data Interchange Standards Association created the Edward A. Guilbert e-Business Professional Award, a lifetime achievement honour bestowed each year in recognition of demonstrating 'outstanding leadership in the e-business standards field'.

7

The ring pull

When: 1963

Where: Dayton, Ohio

Why: It enabled dramatic growth in the soft drinks market

How: The idea for the ring pull came about while Ermal Cleon Fraze was having a picnic and couldn't open his beer can after forgetting his opener

Who: Ermal Cleon Fraze

Fact: After it bought the licence for the new ring pull design in 1963, Pittsburgh's Iron City Brewing Company's sales increased by 233% in a year t's easy to forget what a huge difference a simple invention such as the ring pull has made to everyday life. Prior to its existence, people had to carry special can openers every time they wanted to open a can of beer or coke – which was highly inconvenient at times. While canning food and drink was already popular, accessing the contents of the can was not so straightforward. Fraze's invention has revolutionised the food and drink industry, in particular the sale of beer and soft drinks, adding another level of convenience to the already hugely accessible market.

The background

The idea for the ring pull, sometimes known as the pull-tab, came about out of necessity. Before its invention, cans were opened using implements called churchkeys. However when people inevitably found themselves without such an instrument, they were unable to open their cans. It was while having a picnic in 1959 that Ermal Fraze, from Dayton, Ohio, found himself unable to get into his beer can. He decided there must be a solution to this dilemma, and so set about designing an opener on the can itself. Fraze had some experience in working with metal, including aluminium, which helped a great deal. However he still encountered problems along the way. He had to figure out how to cut an opening into a can top so that it was easy to remove but still robust enough to hold against the can's internal pressure. After spending time designing and testing different models, the solution apparently came to him during a sleepless night – to use the material in the lid itself to form a rivet holding the tab in place.

The idea for the ring pull, sometimes known as the pull-tab, came about out of necessity.

Fraze's first version used a lever that pierced a hole in the can but resulted in sharp and sometimes dangerous edges and led to a number of complaints from customers who had cut fingers and lips on the device. He then set about creating the pull-tab

This billboard was part of an introductory advertising campaign in 1963.

version, which had a ring attached at the rivet for pulling and which could come away from the can completely. The design was patented in 1963 and was sold to aluminium producer Alcoa, and then first licensed to Pittsburgh's Iron City Brewing Company.

The early ring pull detached easily sparking debate over its safety. There was also an increasing amount of environmental protest surrounding the pull tops that were being discarded from cans. Daniel F. Cudzik, of Reynolds Metals, addressed this design flaw in 1975 by developing stay-tabs. His design reduced injuries and also diminished the problem of roadside litter caused by removable tabs. Cudzik's device used a separate tab attached to the upper surface as a lever to weaken a scored part of the lid, which folded underneath the top of the can and out of the way of the opening.

The snap top can was so important to the Pittsburgh Brewing Company's bottom line that it appeared on this 1963 cover of their annual report to stockholders.

Commercial impact

As a result of Fraze's original design, Iron City's sales increased 233% in a year. Another US national giant, Schlitz, also adopted the new tops. However, some brewers were reluctant to embrace the new device. The ring pull tops added 1 to 5 cents in cost to the brewer for each six-pack of cans, which was perhaps one reason for their unwillingness to use them. While they're a mainstay of today's drinks industry, many people thought they were just another fad.

However, it was soon evident that this invention was no passing gimmick, as more beer and beverage companies started to show an interest in the pull tops. By June 1963, 40 brands were using pull-tabs and by 1965, 75% of US breweries had adopted them. Pull tops also started to be used on some oil cans, as well as soup cans and pet food.

For all their convenience, the original pull-tabs created a large amount of controversy, namely for their environmental impact. For 10

Snap tops were so easy. A beautifully manicured woman opens a can without damaging her nails.

years people opened cans by ripping off the pull-tabs and discarding them. The discarded tabs were a danger to wild animals, which suffered death from ingesting the metal pieces. There were also cases of people choking on the tabs, having dropped them into their can by mistake. The stay-on tab solved this dilemma and is the opening device we know today.

It was soon evident that this invention was no passing gimmick, as more beer and beverage companies started to show an interest in the pull-tops. By June 1963, 40 brands were using pull tabs and by 1965, 75% of US breweries had adopted them.

Drinks brands such as Coca Cola have latched on to the ring pull with great enthusiasm. The beverage giant currently distributes 1.6 billion global servings per day, most of them in cans and bottles. The ring pull has transformed the distribution of Coca Cola and other drinks companies – including most beer brands – offering a hassle-free way for consumers to open cans.

What happened next?

With growing debate about environmental damage caused by food and drink packaging, and the proliferation of recycling, you'd be forgiven for voicing concerns over the future of the drink can, which is the most common host of the ring pull. Aluminium, is, however, the most abundant metal in the earth's crust and the aluminium drink can is the world's most recycled packaging container. With this in mind, the ring pull looks set to dominate the drinks industry for years to come.

Since its amendment in 1975 into a stay-on tab, the design of the ring pull has altered very little, as manufacturers and consumers alike remain satisfied with the blueprint. Increasing numbers of food tins are now produced with ring pull tops, to enable people to access the contents with greater ease. And today the device is used on other containers, such as tennis ball tubes. Despite its diminutive size, Fraze's invention permanently revolutionised the commercial food and drink industry.

8

Touch–tone phones

When: 1963

Where: USA

Why: Touch-tone phones changed the way customer service is delivered worldwide, and paved the way for technologies such as Voice over Internet Protocol and the mobile phone

How: Bell Laboratories created a new way of connecting calls by combining two separate tones for each number

Who: Technicians at Bell Labs

Fact: Touch-tone reduced the minimum time it took to dial a 10-digit number from 11.3 seconds to less than a second

Touch-tone was introduced as a faster, more convenient way of placing calls, but its impact stretches way beyond this objective; indeed touch-tone technology has revolutionised the world of business and laid the foundations for today's portable telephones.

The background

Prior to touch-tone, a system called 'pulse dialling' was used to connect customers, built around rotary phones. When a number was dialled on a rotary phone, it would be represented as a series of 'clicks', made by the telephone rapidly interrupting a steady tone, and these clicks would tell the network which number had been dialled.

While extremely reliable, pulse dialling had some frustrating drawbacks. Rotary phones were often bulky and unwieldy, and dialling a number was slow and cumbersome, as the user had to wait for the dial to return to the top before entering a new number. Added to this, long-distance calls required an operator to assist the caller because telegraphic distortion meant that the 'clicks' became jumbled.

Taking note of this, the researchers at Bell Laboratories (the research and development arm of American telecoms company AT&T) began searching for a new standard that eliminated these issues and allowed connections to be performed much faster, as well as improving dialling. Around 1960, they began testing the first iterations of the Dual-Tone Multi-Frequency (DTMF) dialling system.

Rotary phones were often bulky and unwieldy, and dialling a number was slow and cumbersome, as the user had to wait for the dial to return to the top before entering a new number.

The system, which was based on a similar technology used by AT&T's operators to connect customers' long-distance calls, represented each number with a tone comprising a combination of two different frequencies. This meant that connections could be completed much faster, and enabled callers to connect long-distance calls themselves for the first time. It also meant that Bell Labs could finally complete another invention – the push-button telephone, which used separate numbers instead of a dial; Bell Labs had begun researching the push button phone in the 1950s, but had previously lacked the protocol to make it work.

Bell Labs and Western Electric first showcased the touch-tone phone at the World's Fair in Seattle in 1962, and demonstrated it by getting visitors to dial a number on a rotary phone and then again on the new touch-tone keypad and seeing how many seconds they saved. In 1963 the first touch-tone telephone went on sale to the public in Carnegie and Greenberg, Pennsylvania. The Western Electric 1500 featured 10 keys and required a small extra charge to connect to the DTMF system; soon after, a 12-button model was released containing the star (*) and hash (#) keys to provide advanced functionality, and giving the keypad the form it has retained to this day.

Touch Tone, 2500 Set. Ten buttons replace the customary dial on the Bell Telephone Systems new Touch Tone telephone.

Commercial impact

The business world quickly warmed to the technology's ease-of-use and increased speed, as well as the ability to store numbers and switch calls at the touch of a button. Meanwhile, home users discovered several advantages to touch-tone phones – including voicemail, which had never been possible with the old rotary phones.

By 1979, touch-tone had replaced rotary phones as the choice of most users around the world, and in the 1980s AT&T manufactured its last rotary dial phones. By the mid-1990s, they had become a novelty item, the choice of only a few traditionalists who refused to submit to the forces of change. Just two years after the US telephone industry was deregulated in 1984, every home in the country had a touch-tone phone.

Touch-tone technology facilitated the emergence of several new types of phone, notably the cordless phone. At first, it seemed the cordless handset would be nothing more than a fad – sales in the USA rocketed to $850m in 1983, then fell back to $325m in 1984. But the market gradually settled down, and demand began to climb. According to figures from ABI research, sales of cordless phones generated $5.2bn in 2009, a figure that will drop by 20% to $4.3bn in 2014. The research house concluded that the global cordless phone market will contract and be worth around $1bn by 2014, adding that the new digital, broadband-friendly cordless phones will staunch the slow and steady decline somewhat as mobiles and the rise of the smartphone nibble away at its market share.

Home users discovered several advantages to touch-tone phones – including voicemail, which had never been possible with the old rotary phones.

In addition, the development of the touch-tone keypad, with its space-saving combination keys, was a necessary stage towards the development of the mobile phone. Although mobile handsets did not gain a significant commercial foothold until the 1980s, the market has grown out of all proportion since – indeed experts predict that global sales will reach 1.7 billion units in 2011.

More recently, touch-tone technology has expedited the development of Voice over Internet Protocol (VoIP) technology – a system for making phone calls over a broadband internet connection, which emerged around 2004. VoIP systems copied the touch-tone pad, and VoIP technology is backwards-compatible with all touch-tone devices. VoIP is extremely attractive to business because it offers cheap call rates and facilitates conference calls; furthermore, it underpins the technology harnessed by video conferencing services, such as Skype. According to figures from In-Stat, global expenditure on mobile VoIP will exceed $6bn in 2015 – evidence of the huge potential in this market.

Although VoIP is a fairly recent innovation, businesses around the world have been benefiting from touch-tone technology since it first emerged in the 1980s – thanks to the improved customer service experience it offers. Touch-tone's frequency-based dialling system enables callers to communicate directly with a computer using the keypad, meaning that automatic menu selection during calls has become possible. This allows customers to dial in extensions or perform basic functions themselves without the assistance of a receptionist or attendant. Businesses have found that by replacing dedicated staff with these 'automated attendants' they can save huge sums through touch-tone-enabled customer service, whilst being able to deal with a higher volume of calls.

As computers have become more sophisticated, telephone banking services have become possible, where customers can perform actions such as checking their bank balance and transferring money without the assistance of a live person. Along with innovations such as Electronic Funds Transfer, this meant that 24/7 banking was finally a reality, greatly increasing the efficiency of business.

One might assume that, because touch-tone phones handle many of the call-forwarding functions previously assigned to receptionists and PAs, thousands of people in these lines of work would lose their jobs. But so far, it seems this hasn't been the case; in fact, a report from America's National Receptionists Association claimed that the number of receptionists employed across the country had increased from 851,000 in 1980 to 900,000 a decade later. The number of receptionists across the USA now exceeds one million; evidence that, for all the commercial impact of touch-tone technology, it hasn't dented one of the world's biggest job sectors.

What happened next?

Touch-tone is still the dominant technology used for placing calls, and it has directly enabled many other leaps forward in communications technology, perhaps most importantly the invention of the mobile phone. The rotary dial has been seen for years now as a symbol of antiquity, while the push-button keypad has become ubiquitous on communications devices.

Touch-tone will always be credited with streamlining the way we all communicate.

Yet its hegemony faces a sustained challenge from the explosive rise of the internet, which for many people has now replaced the phone as their primary method of communication. Email, instant messaging and social media seem to be partially responsible for the traditional landline's steady decline in recent times, with usage falling in the USA by around 6% a year since 2000 and similar trends being seen across Europe. Indeed, technology has arisen that uses the increasingly redundant landline sockets, such as a lamp that takes advantage of the electricity running though the socket to power itself.

Mobile phones themselves may prove a threat to the DTMF system of connecting calls; smartphones' increasing internet connectivity is rendering the once-essential protocol less so, as calls and messages can be connected over mobile web networks; Skype launched a dedicated Voice-over-IP smartphone in 2008 and many other models now include VoIP applications, in direct competition with the older protocol.

Whether new technologies will prove the death-knell for touch-tone systems remains to be seen, but we need only to look at the phasing out of rotary dial phones to know that even entrenched technology can quickly become obsolete. Nevertheless, touch-tone will always be credited with streamlining the way we all communicate.

9

Equal Opportunities policy

When: 1964, with the passing of the Civil Rights Act

Where: USA

Why: Equal Opportunities policy helped minority groups to acquire opportunities and representation in the workplace

How: Social and political movements

Who: Everyone, from Martin Luther King to striking Ford workers in Dagenham

Fact: On average, women earn only 80% of the salary received by men t was Charles de Montesquieu who observed that 'all men are born equal, but they cannot continue in this equality. Society makes them lose it'. His premise, that inequality is a part of everyday life and is an unavoidable consequence of the social, economic and political system humans create, has been a popular idea that has stubbornly remained in place for over two thousand years – and some might say still exists today.

But most reasonably minded people now accept that inequality does not have to be the 'normal' consequence of life, and that if it continues (especially in the workplace), it will become increasingly unacceptable. Originally associated with female equality, the concept of equality now covers a plethora of circumstances and characteristics, including age, disability, gender, race, religion and belief, sexual orientation and human rights.

The background

As late as the mid-20th century, the world was still a very unequal place – and nowhere was inequality and discrimination more common than in the USA. The country had fought a civil war over the issue of black slavery almost 100 years earlier, but even though the abolitionists had won, and the slaves had been freed, racial segregation remained a fact of life in schools, hospitals, hotels and restaurants around the country.

Sex discrimination was almost as rampant. America's women had not even gained the vote until 1920, and the vast majority of the female population didn't work. Those who did work were grossly under-valued; for every dollar earned by the average man, the typical working woman received around 60 cents.

In the 1950s, the tide finally began to turn. In 1954, the US government overturned the 1896 ruling that legitimised 'separate but equal' racial segregation. In 1955, Rosa Parks, a black woman in Montgomery, Alabama, was arrested for refusing to give up her seat on the bus to a white person; in response, the local community, led by the Reverend Martin Luther King, Jr, launched a bus boycott. King became a key figure in the USA's civil rights movement, orchestrating a series of disruptive but peaceful protests against the country's endemic inequality.

The efforts of King and his followers were rewarded in 1964, when Congress passed the Civil Rights Act – a breakthrough piece of legislation whose ripple effect was to be felt the world over. Originally spearheaded by President John F. Kennedy, and completed by his successor, Lyndon Johnson, the Act made employment discrimination on the basis of sex, race or ethnicity illegal for firms with more than 15 employees, and ended racial segregation in the workplace. Just a year earlier, Congress had passed the Equal Pay Act, which prohibited wage differentials on grounds of gender; on the issue of equality, the USA was now a world leader.

In 1964 ... Congress passed the Civil Rights Act – a breakthrough piece of legislation whose ripple effect was to be felt the world over.

But the rest of the world soon followed suit. In the UK, the first major piece of legislation was the 1970 Equal Pay Act, which was passed in the immediate aftermath of the 1968 Ford sewing machinist strike (recently remembered in the film Made in Dagenham), which erupted when female workers discovered that they were earning up to 15% less than men in the same job. The Ford strike caused the formation of the National Joint Action Committee for Women's Equal Rights, which held an equal pay demonstration in Trafalgar Square in 1969.

The Equal Pay Act came into force in 1975, the same year that the Sex Discrimination Act was passed, which made it unlawful to discriminate on grounds of sex or marital status in recruitment, promotion and training. By then, Britain had already joined the European Union (in 1973). It was thus already obliged to observe Article 119 of the 1957 Treaty of Rome, which had

The infamous 1968 Ford sewing machinist strike.

established that men and women should receive equal pay for equal work. But Britain was determined to show that it meant business and could make big decisions without guidance from Europe. So the Conservative government set up the Equal Opportunities Commission (EOC), whose primary role was to tackle sex discrimination and promote gender equality.

The establishment of the EOC was the harbinger of a string of laws cementing the notion that equality of opportunity should, in future, be the norm for British people. First came the 1976 Race Relations Act, which made it unlawful to discriminate on grounds of race, colour, nationality or ethnic or national origin. Then came the 1983 Equal Pay Act, which amended the 1970 Act to introduce the concept of 'equal value'. The 1995 Disability Discrimination Act made similar provisions to the previous discrimination acts, but also required employers to make 'reasonable adjustments' to premises or working practices to allow a disabled person to be employed. Most recently, the 2010 Equality Act – which consolidated all existing anti-discrimination legislation in the UK – has paved the way for the abolition of a default retirement age, meaning that workers cannot be forced to retire just because they have reached a milestone age.

Commercial impact

Thanks largely to the raft of recent equality legislation, the number of women in full-time employment in the UK has increased by more than a third over the past 25 years, and the story is similar elsewhere; in the USA, for example, almost 60% of employees are women.

Many would argue that the influx of women has brought considerable business benefits. A 2003 research report by the European Commission, entitled 'The Costs and Benefits of Diversity', found that 58% of companies that have implemented diversity programmes have witnessed an improvement in employee motivation. Furthermore, 57% benefited from improved customer satisfaction, and 69% enhanced their brand image.

Although it is hard to quantify the financial benefit yielded by women in the workplace, several academic and professional studies have provided clear evidence. For example a 2004 report from research body Catalyst, based on an examination of diversity policies in 353 Fortune 500 companies, found that higher financial performance was directly proportional to the number of women employed in senior decision-making roles. Furthermore, statistics released in 2006 showed that the influx of female workers contributed more to global GDP growth during the 1980s and 1990s than either new technology or the emergence of global giants China and India.

Women aren't the only ones to benefit from recent equality laws; older workers have benefited too. According to a report published in March 2011, the number of over-65s in work has doubled over the past decade. And, just like the women who have entered the workplace, older workers who have stayed in work have provided genuine financial gains; the aforementioned report stated that over-65s made a net contribution of £40bn to the UK economy in 2010, and this figure is set to reach £77bn by 2030.

A 2004 report ... based on an examination of diversity policies in 353 Fortune 500 companies, found that higher financial performance was directly proportional to the number of women employed in senior decision-making roles.

However, other minority groups have received less benefit from the equality legislation introduced in recent decades. Many believe that the black community, whose struggle for fair treatment catalysed the civil rights movement 50 years ago, has yet to secure genuine parity. This sentiment appears to be borne out by the employment statistics; last January, the Institute for Public Policy Research revealed that almost half of Britain's black population aged between 16 and 24 were unemployed.

Furthermore, many British business leaders believe today's anti-discrimination laws place an excessive burden on businesses at a time when many are struggling to emerge from the maelstrom of recession. In fact the government's own research, released in 2010 ahead of the Equality Act, found that it would cost £190m just to help UK businesses to understand the new legislation.

In addition, hundreds of companies and organisations have seen their profits consumed by legal costs resulting from discrimination cases. These costs have continued to escalate as the equality agenda has branched out into new areas. In 2009, for example, fashion retailer Abercrombie & Fitch was taken to court when 22-year-old shop assistant Riam Dean claimed she had been banished to the stockroom after bosses discovered she had a prosthetic arm that did not fit with their 'beautiful people' policy. Dean, who was essentially claiming discrimination on grounds of physical appearance, won her case and was awarded a total of £9,014, including £136 in damages, £7,800 for injury to her feelings, and £1,077 for loss of earnings.

What happened next?

Clearly, there is still much work to be done before the equality issue is resolved to everyone's satisfaction. While the female workforce has swelled considerably in recent years, the average woman's wage is still only 80% of the average man's; in the finance industry, the gender pay gap stands at a massive 60%. Furthermore, it seems that women are still underrepresented in the boardroom – in 2010, just 12.5% of boardroom positions in FTSE 100 companies were held by women.

The situation seems even more inequitable for other minority groups. Disabled people, for example, still seem some way from fair treatment. Charity Leonard Cheshire recently found that, of the 10 million people in the UK with some form of disability, non-disabled job applicants were twice as likely to receive a job interview as disabled applicants. Discrimination on the grounds of sexual orientation remains an equally sensitive issue; in July 2011, a spokesperson for the London Gay and Lesbian Switchboard said that unfair treatment and persecution were still a reality of working life 'at all levels'.

Clearly, there is still much work to be done before the equality issue is resolved to everyone's satisfaction.

On the other hand, businesses continue to lobby that the law has now moved unfairly against their ability to select the people they want. In particular, they cite regulation from Europe – such as the Employment Equality (Age) Regulations, which came into force in 2006, which ruled employers were unable to use words such as 'energetic' in their recruitment ads because it was ageist, banned employers from asking the age of applicants, and banned them from specifying a set amount of 'experience' on the grounds it could discriminate against younger applicants.

Many employers grumble that they cannot now ask if a job applicant is disabled prior to interviewing them, even if it is, in the real world, totally inappropriate for them to apply. For example, with all the 'reasonable adjustment' in the world, it would be highly unlikely that a fire-fighter could, for example, have sight or mobility problems.

So the road to equality remains a long and rocky one. But, when one considers where we were 50 years ago, it's clear that huge strides have been made since the passing of the Civil Rights Act, and few would argue against the claim that the changes have provided considerable overall benefit to business – and society as a whole.

10
Video conferencing

When: 1964

Where: USA

Why: Instantaneous conferencing tools meant that flying executives round the world for a meeting was no longer necessary

How: AT&T used the World's Fair as a platform to unveil a revolutionary type of telephone that could display picture messages

Who: AT&T

Fact: Experts predict that the value of the video conferencing industry will grow at more than 5% per year between now and 2016

W hen visitors to the 1964 World's Fair wandered into the AT&T pavilion, it must have felt as though they were stepping into the future. Attendees were invited to try out the Picturephone, a strange new form of telephone that displayed a video image of the person on the other end of the line. Given that many homes were still without a television in the mid-1960s, this was a staggering leap forward.

The public had used 'party lines', in which more than two people could contribute to a phone call, for decades, but seeing the person you were speaking to on a screen was an alien concept. Picturephone opened a new frontier in the communications industry, and its legacy lives on in the video conferencing tools that today compete with and complement modern phone conference and Voice over Internet Protocol (VoIP) real-time video tools, such as Skype.

The background

To the majority of the general public, the concept of a phone that could display video images would have seemed fantastical before 1964. However, the technology had been in the pipeline for over 40 years. Bell Laboratories, the research arm of AT&T, first started working on technology to send pictures with phone calls in the 1920s. In 1927, Bell Labs used telephone lines to transmit live television images of US secretary of commerce Herbert Hoover from Washington to New York. The USA wasn't the only country dipping its toe in the water – the German post office developed a rudimentary system, consisting of two closed-circuit television systems linked by cable or radio, during the late 1930s.

The Bell Picturephone, unveiled in 1964.

However it wasn't until the 1950s, following the invention of the transistor and a fall in the prices of cameras and display tubes, that the concept of video technology became viable, and development began in earnest. On 23 August 1955, two California mayors a mile apart spoke to each other via videophone. The following year, Bell Laboratories completed its first system for two-way video conferencing, the Picturephone.

But this early version was riddled with flaws; it broadcast only one image every two minutes, and exceeded the typical telephone bandwidth more than 300-fold. It took Bell and AT&T a further eight years of tweaking and honing before the device, known as Mod 1, was ready for use. The eventual design consisted of a cathode-ray picture tube, a vidicon camera tube, devices for scanning and synchronisation, and a loudspeaker.

Any hope AT&T had of rolling out the Picturephone was snuffed out by its exorbitant cost; a three-minute call from Washington to New York City cost #16.

By the early 1960s America was already waking up to the potential of high-profile, concerted marketing campaigns, and when the Picturephone was finally unveiled in 1964, AT&T plugged and promoted it with everything it had. The company held a grand Picturephone ceremony in Washington, with First Lady Lady Bird Johnson making the first call, to a Bell Labs scientist in New York.

Then came the stall at the World's Fair: AT&T hooked up the Picturephone in its exhibitor pavilion, and created a similar booth at Disneyworld in California. World's Fair attendees were invited to make a video call across the country, and afterwards were quizzed about their experience by marketers.

The problems were apparent as soon as the World's Fair volunteers submitted their responses; they said the controls were baffling and the picture wasn't big enough. Furthermore, any hope AT&T had of rolling out the Picturephone was snuffed out by its exorbitant cost; a three-minute call from Washington to New York City cost $16.

In 1970, AT&T rolled out the Picturephone once more, this time with trials across several cities. The new version, the Mod II, had a bigger screen, and the controls were supposedly more user-friendly. Plus there was a 'zoom feature' and a 12-button touch-tone telephone. But it was still expensive – $125 per month, plus $21 for every minute. Again, the product flopped.

Still, AT&T kept coming. In 1972, AT&T marketers were dreaming of selling three million units and reaping $5bn in revenue by the 1980s, figures quoted in Bell Labs telephone magazine, but it soon became clear that this wouldn't materialise. In the late 1970s, it pushed to market yet again with a new name, the Picturephone Meeting Service – guaranteeing a rather unfortunate acronym. The AT&T marketing team pitched the product at business customers, who would theoretically have more money than general consumers; but still, the product was way beyond the financial reach of the masses.

PictureTel Corp launched a video conferencing system in the 1980s, when ISDN made digital telephony possible. Then, in 1992 AT&T produced the VideoPhone2500, the world's first colour videophone capable of using home phone lines. Costing the small matter of $1,500, the product was pulled within three years, due to lack of sales.

Commercial impact

Today, the video conferencing industry is one of the world's fastest-growing information and communications technology markets. There are estimated to be more than one million telepresence and video conferencing end-points installed around the world – and growing – covering everything from financial institutions to hospitals.

Businesses are increasingly seeking richer ways to communicate with colleagues, customers, and partners and the global economic problems of recent years have only increased interest in and use of video conferencing, as a solution for saving money on business travel and enhancing the work–life balance for workers. It has also created the possibility of virtual businesses, with fellow directors geographically dispersed and interacting through a variety of communications technologies.

Today, the video conferencing industry is one of the world's fastest-growing information and communications technology markets.

And then there are the purported environmental benefits of video conferencing over carbon-emitting travel. 'SMART 2020: Enabling the Low Carbon Economy in the Information Age', a 2008 report from the independent, not-for-profit organisation The Climate Group on behalf of the Global eSustainability Initiative (GeSI), estimated that by 2020 ICT would reduce global carbon emissions by 15%.

Such business benefits have had an enormous impact on the size and value of the market. Global revenues from video conferencing and telepresence systems reached $2.2bn in 2010, growing from $1.86bn in 2009, according to market research company Infonetics. Furthermore, the company predicted in March 2011 that revenues would more than double by 2015, to $5bn. Fellow research firm Gartner went a step further, predicting a market worth $8.6bn by 2015.

Cisco, following the $3bn acquisition of Norwegian video communications company Tandberg, and Polycom are the clear market leaders in the enterprise video conferencing market. Polycom's market strength is at least partly derived from its 2001 acquisition of PictureTel Corp for $362m in stock and cash. At the time, PictureTel's revenues from video conferencing already exceeded $400m. More recently, Polycom bought Hewlett Packard's video conferencing unit within Visual Collaboration business for $89m.

Between them, Cisco and Polycom are reputed to account for approximately 80% of the market. Microsoft, following its $8.5bn acquisition of Skype in May 2011, may well make inroads quickly. Other significant players currently competing for market share in the fast-growing space include Aastra, Avaya, Huawei, Logitech, Sony, and Vidyo, among others.

What happened next?

Perhaps the biggest single reason for the take-off of video conferencing is the falling cost of equipment. The advent of VoIP technology has brought dramatic cost reductions, and many businesses now access video calls via the computer, at a fraction of the price of the old stand-alone systems.

Another key factor is the globalisation of the world's economy. It is now commonplace for a business to have customers all around the world; video conferencing facilitates quick communication with these far-flung business associates, with the personal touch of face-to-face contact.

Many governments are now brokering deals for large-scale adoption of video conferencing; for example, the Spanish government recently finalised a multi-million pound deal with Cisco. And as the market becomes more saturated and the major providers jostle for position, each company is developing bespoke offers, tailoring its packages for small as well as large companies.

Video conferencing facilitates quick communication with ... far-flung business associates, with the personal touch of face-to-face contact.

All these factors have created a communications landscape totally unrecognisable from the one AT&T tried to penetrate back in the 1960s. It would be totally untrue to say that the old Picturephone led seamlessly to today's video conferencing solutions. But, in opening up the traditional telephone call to visual imagery, the Picturephone opened a window onto the future, and gave the world a glimpse of what was possible. We wouldn't have today's billion-dollar video conferencing industry without it.

11

The fax machine

When: 1964–66

Where: USA

Why: Instantaneous communication of key documents transformed the business world

How: Xerox harnessed the potential of telephone lines to create the world's first truly viable facsimile device

Who: Xerox

Fact: The global fax machine market grew by more than 1,300% between 1983 and 1989

Few of the technologies profiled in this book can boast a back-story as lengthy and undulating as that of the fax machine. The first commercial desktop fax machine was the product of more than 120 years of development, featuring some of the world's boldest and most imaginative innovation. Even after such a long incubation period, the technology could not deliver overnight success; the fax machine would not become a staple of office life for another two decades.

Yet, having waited so long for commercial success, it seems the fax machine will be around for a while. In fact, it is estimated that four million new fax machines are still sold each year. The shrieking whirr of a fax transmission may not be so much a feature of office life now, having largely given way to email, but it's still there, and remains a part of the foreseeable future.

The background

Long before the car, the radio and the telephone had even been conceived, inventors were rolling out rudimentary antecedents of the modern fax machine. The first facsimile device was created in the 1840s by a Scottish inventor named Alexander Bain. Bain's machine transmitted electrical signals to a pendulum, which impressed a brown stain on chemically treated paper in response to the stimulation.

Bain's breakthrough sparked the imaginations of inventors of around the world, resulting in decades of innovation. Englishman Frederick Bakewell displayed a working model of a facsimile device at the 1851 Great Exhibition at London's Crystal Palace; 10 years later, Italian Giovanni Caselli invented the pantelegraph, the world's first commercial facsimile device. By the 1920s, American Dr Arthur Korn was firing images across continents using his photoelectronic scanning system. In 1922 Korn sent a picture of the Pope from Rome, which appeared on the front page of the New York Times just a few hours later.

The fax failed to make the leap from scientific wonder to everyday business reality ... because of the practical drawbacks of the cumbersome and expensive early machines ... Perhaps most importantly, there was no real gap in the market.

But, despite all these advances, the fax failed to make the leap from scientific wonder to everyday business reality. Partly, this was because of the practical drawbacks of the cumbersome and expensive early machines. Another factor was the attitude of the authorities: America's Federal Communications

Commission (FCC) didn't authorise the development of fax machines on a commercial basis until 1948. And, perhaps most importantly, there was no real gap in the market. In an age before globalisation, the average business was distinctly parochial in its outlook, and few companies had any real need for a machine that could transmit documents across vast distances in minutes.

But then, in the 1960s, everything began to change. Xerox, a burgeoning office equipment manufacturer that had already experienced huge success with the world's first automatic office photocopier, decided to try something even more ambitious – a fax machine that would be cost-effective for everyday businesses.

The company had been experimenting with 'xerography', a process that harnessed light and electricity to impress an indelible image on paper, for more than two decades, but had been held back by a lack of suitable transmission systems. Then, in 1961, communications giant AT&T unveiled Telpak – a low-cost transmission service channelled through wideband. Crucially, this imbued Xerox with the belief that the fax could become a viable, accessible business solution.

Xerox rolled out the office fax machine in two key stages. First, in 1964, the company introduced the concept of Long Distance Xerography, or LDX. This technology, the product of three years' painstaking research and development, relied on a document scanner to send the image, a broadband transmission link to transport it and a document printer to receive it. The image was read by the scanner before being coded into electronic signals and sent via the broadband conduit.

LDX was revolutionary but, like all the fax devices which preceded it, it failed the commercial viability test; crucially, the technology underpinning xerography was not compatible with the telephone lines of the time. Yet Xerox, having come so far, was determined to press on and complete its mission.

In April 1966, Xerox teamed up with the Magnavox Company to develop the Xerox MagnafaxTelecopier, a desktop fax machine which, finally, harnessed conventional telephone lines. Weighing in at just 46lbs, a fraction of the weight of its predecessors, the Telecopier was connected to telephone circuits with an acoustic coupler – a relatively simple advance worth billions of dollars. The Telecopier was capable of sending a full-page message in six minutes, and a verification signature in a staggering 30 seconds.

In April 1966, Xerox teamed up with the Magnavox Company to develop the Xerox MagnafaxTelecopier.

Commercial impact

By the time the Telecopier was ready for market, Xerox had been pushing fax technology for more than five years – plenty of time for its marketing team to devise a decent strategy. The Xerox marketeers honed in on a typical user, the busy office administrator who relied on quick and accurate receipt of information, and tailored their promotional messaging accordingly. The strategy brought instant success. By 1968, Xerox had broken away from Magnavox and was marketing its own Telecopier, and nearly 30,000 fax machines were in use across the USA by the early 1970s.

Yet, in the context of the worldwide office equipment market, this was a drop in the ocean. And still the fax machine had to wait for its period of global pre-eminence. As new contenders, such as Exxon, entered the market, each with its own method of using telephone lines, standards became muddled and many fax machines were simply unable to communicate. Furthermore, as former Muratec and 3M manager Michael Smart said, 'the machines were big, noisy and gave off fumes'.

In 1983 ... the Group 3 protocol ... laid down a single standard for the transmission of faxes. Now, every fax machine would communicate on the same parameters; the golden age of the fax machine had well and truly begun.

At the end of the 1970s, the fax was still a relative rarity. But the wheels of innovation kept turning – by this time the drivers were based in Japan, a country which saw the fax as a catalyst for its burgeoning economy and a convenient conduit for its complex pictorial alphabet. In 1983, the international communications body, CCITT, introduced the Group 3 protocol, which laid down a single standard for the transmission of faxes. Now, every fax machine would communicate on the same parameters; the golden age of the fax machine had well and truly begun.

Between 1983 and 1989, the number of fax machines in use worldwide soared from 300,000 to over four million. By the end of the decade, Michael Smart says, 'it took closer to 10 seconds, rather than the former six minutes, to transmit each sheet or document'. All around the world, offices were humming to that squealing beat.

In fact some business people argue that the fax machine had a greater impact on the business world than email. Certainly it enabled almost instant communication multinationally, facilitating substantial growth in international trade.

What happened next?

The fax machine didn't dominate the office landscape for long; by the late 1990s, more and more offices were getting hooked up to the internet, and using email to send documents which previously had to be sent in physical form. The fax machine, having waited over a century for its period of pre-eminence, had been bypassed in little more than a decade.

Today, given the predominance of email, it's easy to write off the fax machine as an obsolescent relic of the 1980s, trapped in its time, like Walkmans and shoulder pads. Why, one might ask, would anyone need an expensive, high-pitched, stand-alone contraption to do the work of an email, which can zoom around the world in nanoseconds – and take scanned documents with it?

Well, actually, many people still have such a need. Many countries, and professionals, refuse to accept an electronic signature on a contract, so faxed signatures are still necessary. The fax machine is also capable of transmitting handwritten notes, which cannot be sent by email. Finally, and perhaps most importantly, faxes are less vulnerable to interception than emails – a major advantage for people looking to send sensitive information over long distances.

Far from being beaten down by the email, fax manufacturers have rolled with the punches. For example fax over IP, or 'FoIP', allows a user to send a message from email or internet to a capable fax server. Android phones carry fax functionality, and Windows 7 now comes with fax-sending software. Many businesses now harbour a fax server, which can create an electronic copy of an incoming fax and forward it to the company's employees on paper or email, and various manufacturers offer multi-function units, combining the fax with other everyday office equipment.

Many people still have a need [for the fax machine]. Many countries, and professionals, refuse to accept an electronic signature on a contract, so faxed signatures are still necessary.

These advantages and attributes have enabled the fax machine to retain a relevant place in the commercial world. In April 2011, a CBS report claimed that the USA alone will spend $55m on new machines over the course of this year. As more and more people who once relied on fax machines retire and are replaced in decision-making positions by those who have only ever known email, the picture may change; but for now, the fax machine is clinging on.

12

The plastic bag

When: 1965

Where: Sweden

Why: The durability and affordability of the bags has seen them become an essential part of the supermarket industry

How: The idea to produce a simple strong plastic bag with a high load-carrying capacity was patented in 1965 for packaging company Celloplast of Norrköping

Who: Sten Gustaf Thulin

Fact: Around one million plastic bags are used every minute worldwide

Since their inception in the late 1960s, plastic bags have revolutionised everyday life around the world. Made from polyethylene, they have become the most common type of shopping bag, helping millions of people to carry items from the shops to their homes, while eradicating the age-old dilemma of splitting paper bags. However, in recent years this ground-breaking convenience tool has sparked much debate surrounding its environmental impact. The question of what to do with plastic bags once they've been used hovers like a dark, non-degradable cloud. Each year, millions of discarded bags end up as litter that will take up to 1,000 years to decompose. Campaigns such as 'Banish the Bags' and 'I'm Not a Plastic Bag' have been launched around the globe in an attempt to solve this growing problem.

The background

While the history of plastics goes back more than 100 years, plastic bags weren't properly invented until the 1960s. Patent applications actually date back to the 1950s, when the first plastic sandwich bag was produced. However, the lightweight shopping bag we know today was the brain-child of Swedish engineer Sten Gustaf Thulin, who discovered that you could make bags from polyethylene – which is made up of long chains of ethylene monomers – by a process called blown film intrusion. He developed the idea for creating a simple bag out of a flat tube of plastic for Swedish packaging company Celloplast, which it patented in 1965, giving Celloplast a virtual monopoly on plastic bag production. The company expanded and built manufacturing plants across Europe and the USA. However, it wasn't long before other manufacturers realised the potential for polyethylene bags, and the patent was successfully overturned in 1977 by US petrochemicals group Mobil. This led to a number of firms exploiting the opportunity to introduce convenient bags to all large shopping stores.

While the history of plastics goes back more than 100 years, plastic bags weren't properly invented until the 1960s.

As of 1969, bread and other produce was sold in plastic bags, and New York City became the first city to collect waste in larger plastic bags. It wasn't until a few years later that plastic carrier bags became a mainstay of the retail sector. In 1974 US retailing giants such as Sears and J. C. Penney made the switch to plastic merchandising bags, and the supermarket industry was introduced to these revolutionary carriers in 1977, as an alternative to paper bags. However, they weren't properly incorporated into supermarkets until 1982, when two of America's largest grocery stores, Kroger and Safeway, replaced traditional paper sacks with the polyethylene bags. News spread across the pond, and British retailers latched onto the idea, introducing these convenient bags to their stores.

Commercial impact

Plastic bags have unquestionably had an enormous impact on today's world. Since the 1980s, households around the globe have come to rely heavily on these convenience items to transport goods effectively. When you consider that each year we consume between 500 billion and one trillion plastic bags worldwide, the production of polyethylene bags has certainly been big business. In the early 1980s, McBride's Dixie Bag Company, Houston Poly Bag and Capitol Poly were all instrumental in the manufacturing and marketing of plastic bags in the USA, which led to Kroger and Safeway supermarkets introducing them into their stores.

The proliferation of plastic bags has, however, resulted in grave problems for the environment. Polyethylene takes years to degrade – depending on its density, it can take a single plastic bag anywhere from 20 to 1,000 years to decompose, and when they do, they excrete toxic substances into the earth

around them which are harmful to wildlife. The question over what to do with used poly bags is mired in controversy as environmentalists strive to reduce the enormous amount of waste they create. One such solution has been the development of a sturdier, re-usable plastic bag – often known as a 'Bag for Life' – that consumers can use many times over, rather than using new bags every time they go shopping. Most supermarkets now have these bags on sale in the UK, which has gone a little way to reversing the commercial success of lightweight traditional plastic bags, although they're still fairly prevalent throughout the world.

The original Waitrose 'Bag for Life'.

The proliferation of plastic bags has resulted in grave problems for the environment ... It can take a single plastic bag anywhere from 20 to 1,000 years to decompose, and ... they excrete toxic substances into the earth around them.

What happened next?

The controversy surrounding plastic bags has given rise to many national and international campaigns to reduce their global consumption. Devastating statistics, and images of wildlife that has been harmed by litter – predominantly plastic bags – have been plastered across the media in an attempt to persuade people to stop using them in such numbers. In 2006, the United Nations released some alarming figures, including the fact that 10% of the plastic produced every year worldwide winds up in the ocean, 70% of it settling on the ocean floor, where it is unlikely to ever degrade. The impact this is undoubtedly having on nature is overwhelming.

Various strategies have been suggested over the years to alleviate the problem. While some advocate the banning of plastic bags and a return to paper bags, or more heavy-duty bags, others believe that re-use and recycling of old bags is the best solution to the dilemma. In Belgium, Italy and Ireland, legislation has been introduced to discourage the use of, and encourage the recycling of, polyethylene bags. A plastic bag levy was introduced in Ireland in 2001, which led to an approximate reduction of more than 90% in the issuing of plastic shopping bags. China and some US states have also introduced taxation on plastic bag consumption.

In Belgium, Italy and Ireland, legislation has been introduced to discourage the use of ... polyethylene bags. A plastic bag levy was introduced in Ireland in 2001, which led to an approximate reduction of more than 90% in the issuing of plastic shopping bags.

In Britain, plastic manufacturers have pledged to make bags lighter and thinner, and to use more recycled plastic, which they claim will reduce the environmental 'footprint' of poly bags by 50% over the next few years. However, the industry is trying to head off moves by politicians to support a total ban on free throwaway bags, in favour of re-usable alternatives. Marks & Spencer has led efforts by retailers to crack down on plastic bag waste, by introducing a 5p charge to customers in 2009 for using food carriers. Since then, the company has seen a reduction of 80% in the number of bags it hands out.

Whatever the future holds for plastic carrier bags, they've certainly made headlines in their lifetime, for all manner of reasons. An all-out global ban is impossible to imagine. However it's apparent that a change in the attitude of consumers is required in order to reduce the environmental impact of this once-loved creation.

13

The microwave oven

When: 1965

Where: USA

Why: Created a new market of instant, microwave-friendly foods

How: Experiments with radar equipment during World War II led to an accidental discovery that ultimately changed the world's cooking habits.

Who: Raytheon's Dr Percy Spencer

Fact: By 2009, 93% of UK households owned a microwave oven

Nobody set out to invent a microwave oven, or even suspected that there could be an electronic alternative to conventional solid fuel, gas or electric radiation for domestic or commercial cooking. It was a happy accident. Now it is an essential item in almost every kitchen in the developed world. Many products, particularly in the electrical field, came about as a by-product of research in another field. War is said to accelerate innovation, and World War II technology definitely had its part to play in the invention of the microwave oven, though indirectly.

The background

Microwaves are electromagnetic impulses with a wavelength roughly between 30cm and 1m (or between 1 and 300 GHz for the technical). They are really just very high frequency radio waves. Their existence was predicted in 1864 by James Clerk Maxwell and first demonstrated in an apparatus put together by Heinrich Hertz in 1888.

Though Bell Telephone Laboratories in the USA had observed in 1934 that a high voltage, high energy field would have the effect of heating 'dielectric' materials (materials which are poor conductors of electricity but good at supporting an electrostatic field), nobody thought of applying them to food. During World War II, however, microwaves became extremely important because these wavelengths were the most effective for radar. Soon after the outbreak of war in 1939, a research team at Birmingham University was asked to develop a radio wave generator operating at a wavelength of about 10cm.

Within a year, two members of the team, John Randall and Harry Boot, developed what is known as a cavity magnetron. It was a great success but needed urgent refinement, so great was the demand for radar installations. Development was transferred to the GEC Research Laboratories in Wembley, which had been turned over to UK government work at the beginning of the war. The early radar sets had to be produced there because there wasn't time to tool up a factory.

One of the early GEC magnetrons was flown to the USA and demonstrated there by Sir Henry Tizard. It was just one of a number of technical innovations brought to the USA in the hopes of securing assistance in maintaining the war effort. The official historian of the Office of Scientific Research and Development, James Phinney Baxter III, said later that the Tizard Mission to America in 1940 'carried with it the most valuable cargo ever brought to our shores'. Though the mission carried details relating to the jet engine and the atomic bomb, among other key innovations, he wasn't referring to them but to Magnetron 12.

Spencer [noticed] when standing close to the magnetron ... that a chocolate and peanut bar he happened to have in his pocket had partially melted.

An engineer from the vacuum tube manufacturer Raytheon, Percy Spencer, attended the demonstration and managed to convince Tizard to award Raytheon a low-volume contract. By the end of the war, thanks to the improved manufacturing methods introduced by Spencer, Raytheon was making 80% of the world supply of magnetrons for radar systems.

Toward the end of the war, in 1944, after Raytheon had become a major defence contractor, Spencer was testing one such system when, standing close to the magnetron, he noticed that a chocolate and peanut bar he happened to have in his pocket had partially melted. The next morning Spencer placed an egg near the magnetron tube. It exploded after a few minutes. He then popped some corn, realising that if low-density microwave energy leaking from a radar device could cook a nearby egg, perhaps it could be harnessed to cook other foods.

Raytheon set a team to work on Spencer's idea, and on 8 October 1945 filed a US patent for the cooking process. The principle whereby microwave radiation causes molecules common in food – especially water and fat molecules – to rotate, generating heat that is then transmitted to the rest of the mass, was already understood. But work needed to be done on containing the radiation within the box and distributing it evenly through the food. Microwaves do not cook from the inside out, as many people believe, though they do distribute heat more efficiently than conventional ovens. How deep microwaves penetrate depends how the food is composed and their frequency, with longer wavelengths penetrating deeper than shorter ones.

At the same time an oven was placed in a Boston restaurant for testing. Development accelerated once the war ended, and in 1947 the company built the first commercial microwave oven in the world, called the Radarange to proclaim its heritage. The water-cooled 3kW oven was six feet tall, weighed 340kg and went on sale at about $5,000.

Dr Percy Spencer introduces Raytheon's new invention, the first commercial microwave.

Commercial impact

Early commercialisation was slow and it was not until 1965, when Raytheon acquired Amana Refrigeration, an Iowa-based manufacturer with a well-established distribution channel, that the domestic market opened up. The first household model was also branded Radarange and sold for $495 – too expensive for a mass market, but over the late 1960s the domestic microwave oven was established as a niche item, helped by the fact that its role in freeing women from household drudgery put it in the same class as the washing machine and the dishwasher.

One restricting factor was the fear of microwaves, which were, quite wrongly, associated with radiation. Nonetheless, by 1975 sales of microwave ovens exceeded gas cookers for the first time. By 1976, 17% of all homes in Japan were doing their cooking by microwave, and in the same year the microwave oven became a more commonly owned kitchen appliance than the dishwasher, reaching nearly 52 million US households, or 60%.

The first household model [was] too expensive for a mass market, but over the late 1960s the domestic microwave oven was established as a niche item.

What happened next?

Improved magnetrons were developed in Japan and elsewhere and, as every white goods manufacturer put out a range of microwave ovens, the price dropped to commodity levels. Options and features were added, in particular, supplementary convection heating or grilling to add to the oven's versatility and enable it to simulate traditional roasting and baking functions. The tendency of a static microwave to cook patchily was overcome by the addition of rotating tables, which are now standard.

Now food preservation technology is being developed to extend shelf life and improve food quality and nutrition. Microwave Sterilisation Process technology immerses packaged food in pressurised hot water while simultaneously heating it with microwaves at a frequency of 915 MHz – a frequency that penetrates food more deeply than the 2450 MHz used in home microwave ovens. Whether popular or not, microwave ovens are set to remain a feature in homes and offices the world over.

14

The smoke alarm

When: 1965

Where: USA

Why: The smoke alarm has revolutionised safety alert systems and saved lives

How: Adding batteries and an audible alarm to early conceptual work made the smoke alarm suitable for the residential market, and therefore for mass consumption

Who: Duane Pearsall and Stanley Peterson

Fact: Many modern smoke alarms use the same type of radioactive material as is used in the space programme

This palm-sized life-saver is discreetly located in the majority of residential and commercial properties today. Tucked neatly out of plain view – generally on walls above eye level, or on ceilings, the smoke alarm or smoke detector is one modern safety device that no home, office, restaurant, hotel or establishment should be without.

Fire detection has changed radically over the years – 200 years ago the method of alerting of the danger of fire involved banging on doors, blowing whistles, ringing church bells and even shooting into the air with a gun. Thankfully, times have changed, and in the lifespan of the humble smoke alarm countless lives have been saved. As inventors develop and hone the device to produce updated models even more effective in detection, the smoke alarm's invaluable work looks set to continue.

The background

The history of the smoke alarm began with the US inventor Francis Thomas Upton, who created the first fire alarm in 1890. The patent number for this first device was 436,961. However, the invention was very basic and little interest was shown in the product. Upton's model was soon replaced by new models, the next one of which was created through a lucky accident.

The succeeding smoke alarm was invented by chance in 1930 by the Swiss physicist Walter Jaeger. Jaeger was attempting to invent a sensor that would detect poisonous gas. His idea was that the gas would enter the sensor and ionise the air, causing the meter current to be altered. However, the gas was ineffective in registering in the current meter on the device. Jaeger, believing that he had failed, took a cigarette break. Not long after lighting his cigarette, Jaeger noticed that the meter current on the detector had dropped and he discovered that it was smoke that the sensors had detected, not gas. This breakthrough moment revolutionised the future invention of smoke alarms. Nevertheless, it was not until more than a quarter of a century later that smoke alarms were developed for the consumer market.

While these early devices hardly made a commercial impact, the scientific ground work conducted by Upton and Jaeger paved the way for successive smoke alarm inventors.

Not long after lighting his cigarette, Jaeger noticed that the meter current on the detector had dropped and he discovered that it was smoke that the sensors had detected, not gas.

Manufacturing remained very expensive for the next three decades and it wasn't until the mid-1960s that affordable smoke alarms could be mass produced. The first few smoke alarms that were available were so expensive that only large companies could afford to install them.

The first battery-powered smoke alarm was developed by US inventors Duane Pearsall and Stanley Peterson in 1965. Their design was operated by batteries, which made it cheaper and easier for home owners to run them, and from the late 1960s onwards, smoke alarms were produced for the residential property market. However, Pearsall and Peterson soon noticed a flaw in their first model. The original alarm contained large batteries that were expensive and impractical to change regularly, which led to the invention of a model that made use of smaller, AA batteries, making it more commercially viable. Both of these models were made of metal and were resistant to fire. The battery was made by Gates Energy and it wasn't long before the heavy-duty rechargeable batteries were replaced by single-use AA batteries.

The devices proved very successful. In 1975 Peterson and Pearsall began mass producing their devices under the company name Statitrol Corporation and became the first in the history of smoke alarms to sell the product commercially. By 1977, the company was shipping 500 units daily.

However, in 1980 Statitrol had sold its invention rights to electronic goods manufacturing giant Emerson Electrics. Later, the US manufacturer Sears would end up selling the product, which remains popular today.

Commercial impact

The audible alarm quickly emerged as the most popular type of device. The alarm sends out a high-pitched beeping or ringing noise, and is activated when smoke is detected, thus causing a fire alert. However, consumer demand and a drive for difference has led to the evolution of the simple smoke alarm and now a plethora of alarms can be purchased that are specifically designed for the needs of the hard of hearing or the visually impaired, which use strobe lights or a vibrating action to create the alert.

Many modern smoke alarms use the same type of radioactive material to detect smoke particles as the silver foil used in the US space programme. This type of smoke detector is known as an ionisation chamber smoke detector (ICSD) and is quicker at sensing flaming fires that produce little smoke. The ICSD was invented in Switzerland in the early 1940s and was introduced into the USA in 1951. It uses a radioactive material to ionise the air in a sensing chamber; the presence of smoke affects the flow of the ions between a pair of electrodes, which triggers the alarm. However, use of this material been questioned in recent years because concerns have been raised about the long-term health implications of exposure to a radioactive biohazard.

Although the levels of americium-241 present in most home smoke detectors are considered to be safe, the possibility of long-term risks has led to the development of photoelectric detectors, which contain no radiation and are in fact more affordable. Photoelectric detectors use an optical beam to search for smoke. When smoke particles cloud the beam, a photoelectric cell senses the decrease in light intensity, which activates the alarm. This type of detector reacts most quickly to smouldering fires that release relatively large amounts of smoke, and has proved to be a popular alternative.

Many modern smoke alarms use the same type of radioactive material to detect smoke particles as the silver foil used in the US space programme.

What happened next?

The smoke alarm is one invention that has stood the test of time. It is now a legal requirement for all commercial buildings and rented accommodation in the majority of developed countries to be fitted with a fully functioning smoke alarm. This is particularly important as the electronic equipment we use on a daily basis increases. In the USA, 96% of homes have at least one smoke alarm, according to the National Fire Protection Association, but many people fail to test that their alarm is working regularly or to change the batteries. According to the London Fire Brigade, while 85% of UK homes have an alarm fitted, many fire-related deaths are caused by people being overcome by smoke and fumes, not just by burns, which is why newer models are being developed that detect the presence of smoke at an earlier stage.

The smoke alarm has undergone several transformations and has come a long way since its accidental beginnings. While many of the earlier smoke alarm models were prone to giving false alarms, models today have developed and come with features that aim to reduce the occurrence of false alarms.

The smoke alarm is one invention that has stood the test of time. It is now a legal requirement for all commercial buildings and rented accommodation in the majority of developed countries to be fitted with a fully functioning smoke alarm.

However, as a health and safety device the models we use today are in a constant state of evolution, as product inventors and designers continue to experiment to create prototypes that are even more effective in detecting smoke and saving lives.

Following in the footsteps of Upton is an award-winning young British teenager called James Popper, who came up with an idea for an infrared detector that identifies fires in their infancy. Popper had the idea after a fire destroyed the kitchen of a family friend, leading him to invent the CookerSmart IR Kitchen Flame Detector, which reads the specific infrared frequency bands of flame. While this model of alarm is still relatively new to the market, it appears that it won't be long before new technologies take its place.

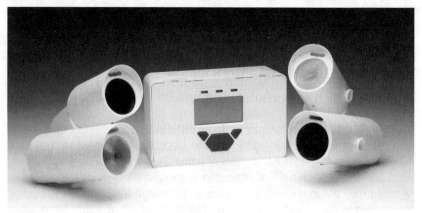

The Fireray 3000 end-to-end optical beam smoke detector provides a new solution for commercial buildings.

15
Kevlar

When: 1965

Where: USA

Why: The product transformed the security industry

How: A DuPont chemist stumbled across the fibre when researching ways to create a new polymer for tyres

Who: Stephanie Kwolek

Fact: Kevlar is five times stronger than steel, based on equivalent weights

I f you've ever been employed to protect others, there's a good chance you'll have been protected by Kevlar. The world's toughest fibre, commonly used in the manufacture of body armour, helmets and bulletproof vests, turned 46 in 2011. While many of those who wear it would be thinking about leaving the firing line and putting their feet up around this age, Kevlar is still going strong, helping the world's troops, firemen and policemen to defuse high-pressure situations.

Kevlar is also widely used in the construction of tornado shelters, bridge suspension cables, brake pads, space vehicles and loudspeaker cones. It was even used in the highly publicised (and much criticised) roof of Montreal's Olympic Stadium, and the adidas F50 boots worn by some of the world's best-known soccer players. Not bad for a product that was invented by mistake.

Background

The history of Kevlar can be traced back to 1927, when science and research giant DuPont created a $20,000 budget for 'fundamental research work': the sort of work that may not lead directly to the development of a new product, but may create a foundation for subsequent discoveries – or just prove interesting for its own sake.

In fact, the group set up to carry out the fundamental investigation soon gained an understanding of how to condense polymers, a process that led to the creation of extremely durable, adaptable materials. The invention of nylon followed in 1938, as did the invention of a string of commercially successful brands, including Teflon and Stainmaster.

The material Kwolek and her technician had created was stiffer and stronger than any fibre previously created [and they] soon realised that they'd stumbled upon a truly unique discovery.

Yet perhaps the most crucial offshoot of DuPont's polymer research was Kevlar, which was invented by chemist Stephanie Kwolek, by accident, in the early 1960s. Kwolek was trying to create a thick, viscous polymer to use in tyres. However, during her research she found that her polymer was becoming cloudy and runny – hardly the solution she was after.

Nonetheless, Kwolek and her technician spun the polymer into a fibre. Better to try it than just throw it away, they thought. Amazingly, when it went through the spinneret, it didn't break. In fact, almost nothing could break it.

Stephanie Kwolek at the Pioneering Research Lab at the Experimental Station, preparing a polymerisation experiment.

The material Kwolek and her technician had created was stiffer and stronger than any fibre previously created. Better still, Kwolek's miracle product didn't break or wither under pressure; the fibres were woven together as tightly as a spider's web, meaning that it was almost impermeable. When they tried burning it, it wouldn't melt. When they tried freezing it, it didn't become brittle and snap. Various different chemicals had no effect. Kwolek and her technician soon realised that they'd stumbled on a truly unique discovery.

The product they'd created was strong and resilient, but also light and elastic; tough enough to be used in military defence and springy enough to be used for sports. DuPont introduced its new development 1965, but did not produce it in large quantities until 1971, and it was later that decade that commercial products were introduced onto the market.

Commercial impact

It's almost impossible to put a figure on the commercial value of Kevlar, since the fibre is so widely used in so many significant applications.

It took DuPont six years to bring the product to market, but its potential soon became clear, particularly in the civil and military defence sectors.

The commercial impact, however, quickly became apparent. Following its creation, DuPont asked its Pioneering Lab to turn the marvellous new polymer into a commercial money-spinner. The possibilities seemed – and remain – almost endless. In addition to bulletproof vests, which have come to define the product, DuPont started to develop Kevlar for radial tyres, brake pads, racing sails, the cables of bridges and the shells of spacecraft. Researchers even found that the material could be used in loudspeaker cones and marching snare drums; with each passing year came new possibilities, and new commercial avenues to explore.

It took DuPont six years to bring the product to market, but its potential soon became clear, particularly in the civil and military defence sectors. Kevlar's commercial prospects were further boosted when America's National Institute of Justice commissioned a four-phase programme to develop bulletproof clothing using the new material. By 1973, researchers at the US army's Edgewood Arsenal had developed a bulletproof vest, and by 1976 the product was ready for use by US police officers in the field, with the US military adopting it for flak jackets in 1978. Demand has not abated since. In 2011 global government spending on military body armour was set to reach $1.19bn, according to the report 'The Military Body Armour & Protective Gear Market 2011–2021'. Vector Strategy forecast in 2009 that the US military would procure body armour to the value of $6bn between 2009 and 2015.

BCC Research, in a separate report, estimated that the global market for advanced protective gear and armour was worth $4bn in 2010, and would rise to $5.2bn by 2015. However, BCC's report encompassed ancillary components (gloves, headwear and respirators); chemical, biological, radiological and nuclear gear; thermal protective; and armour and bullet resistant products, with ancillary components alone accounting for around 60% of the market's value.

Since the original discovery of Kevlar, DuPont has refined the product into several distinct variants. Notable varieties include Kevlar K29, the branch that is used for body protection; Kevlar49, typically used in cable and rope products; Kevlar K129, a high-tenacity variant typically used for ballistic applications; and Kevlar KM2, which is most commonly found in armoured vehicles.

Although a similar fibre called Twaron was launched commercially in 1987, having been developed in the 1970s by Dutch company ENKO (now part of AKZO), Kevlar has retained a huge market share, and demand for the product has remained high. In 2010 DuPont's sales were reported to have reached $31.5bn, with $3.4bn generated by its safety and protection unit, within which textile apparels – namely Kevlar and Nomex – reportedly account for 20% of sales, although DuPont refuses to divulge exact figures.

What happened next?

New inventions based on Kevlar technology continue to enter the market; indeed, the material is now being used to reinforce sheds and bunkers, as tests have shown that it can repel projectiles flying at speeds of up to 250mph. In November 2010, an anti-grenade net made up of tiny Kevlar fibres was launched by Qinetic – a further example of Kevlar's continuing importance to civil and military defence forces.

New inventions based on Kevlar technology continue to enter the market.

Buoyed by this continued innovation, DuPont has constructed a new $500m Kevlar production facility near Charleston, South Carolina. It is believed that the new Cooper River Kevlar start-up plant will increase global Kevlar production by 25% and double sales in developing countries by 2015. Brazil, for example, represents a major growth opportunity, due to the number of murders in the country, which total around 40,000 each year.

Kevlar's market dominance is constantly under threat from competitors, who spend millions trying to develop competitive rivals to the product. International trade agreements also pose a threat; for example, the Korea–USA Free Trade Agreement could ultimately tempt US consumers into importing cheap Korean Kevlar substitutes. However, there is no doubt that, at present, Kevlar is still far and away the leading player in its sector.

Surprisingly, Stephanie Kwolek has played a relatively minor role in the development of Kevlar since she first chanced upon the product. She remained a prolific inventor for the rest of her working life and went on to achieve 28 patents during her research career. She is now a part-time consultant for DuPont, and a mentor for young female scientists.

In recognition of her efforts, Kwolek has received the prestigious Kilby Award and the National Medal of Technology. In 1999, at San Franciso's Exploratorium, she was given the Lemelson-MIT Lifetime Achievement Award – an accolade that recognised her pioneering role in mentoring young female scientists, as well as her own research efforts.

16

Aerosol deodorant

When: 1967

Where: USA

Why: The ease of use of the aerosol deodorant finally provided a solution to the age-old problem of body odour

How: Two cosmetics giants went into battle, and gave birth to the world's first aerosol antiperspirant deodorant

Who: Gillette, Carter-Wallace

Fact: Analysts predict that the global deodorant market will be worth $12.6bn (£7.7bn) by 2015

Every product mentioned in this book has met a clear market need, but arguably none has tackled a more significant need than the aerosol deodorant. Body odour has blighted humanity since the dawn of civilisation, and it's taken us thousands of years to find a solution. The early Egyptians drew scented baths to mask the smell of dirt and sweat; the Greeks and Romans used perfume; ancient Middle Eastern societies even relied on full body hair removal to keep bad smells away.

The aerosol deodorant, which came onto the market in the 1960s, was as significant a breakthrough as any previously made in the long battle against body odour. The spray-on solutions rolled out almost 50 years ago brought new levels of user-friendliness and enabled manufacturers to bring out deodorants with antiperspirant properties, greatly improving overall efficacy.

The pre-eminence of the aerosol deodorant was relatively short lived. Concerns about the environment, and the emergence of alternative methods of application, eroded consumer confidence and ate into market share. But, despite the fierce criticism and competition, aerosols continue to constitute a substantial share of the deodorant market, which is worth half a billion pounds a year in the UK alone.

The background

It is thought that underarm deodorants were invented as long ago as the 9th century, but development did not begin in earnest until the late 1800s, when scientists discovered that two glands were principally responsible for the production of sweat. In the 1880s, a team of chemists in Philadelphia developed a rudimentary paste for the regulation of body odour. Although the product, branded Mum, was messy and difficult to apply, the scientists knew they were on to a winner.

In 1903 the world's first antiperspirant deodorant, EverDry, came onto the market, using aluminium chlorohydrate (ACH) salt to block the secretion of sweat. Meanwhile, development of Mum continued; in the late 1940s, a woman named Helen Barnett Diserens created an underarm product based on the design of a ball-point pen, heralding the birth of roll-on deodorant.

By the 1950s, odour-combatants accounted for a crucial chunk of the toiletries market. However, the sector was still divided into two contrasting but complementary segments: deodorants, which tackle the bacteria that cause body odour, and antiperspirants, which inhibit the glands that secrete sweat. As the market began to boom, cosmetics giants began searching for a product that could combine the best of both worlds.

At the same time, aerosol technology was developing fast. In 1926 a Norwegian chemical engineer, Eric Rotheim, invented a refillable can, complete with valve and propellant, to help him wax his skis. The idea behind the aerosol was relatively simple – a pressurised fluid was used to push another fluid up through a tube, and out through a nozzle at the top.

In 1928, Rotheim sold the patent to America, and the aerosol gradually began to gain traction. In 1939, American Julian S. Kahn received a patent for a disposable spray can. Two years later, Americans Lyle Goodhue and William Sullivan came up with the first refillable spray can, known as the 'bug bomb', using compressed gas.

During World War II, US soldiers began using aerosols for insecticide, and the patent for the product was released for commercial use once the conflict was over. By the 1950s a range of beauty products, including perfume and shaving creams, were being dispensed from aerosol cans.

Buoyed by these inroads, scientists began researching ways of developing an aerosol deodorant that could also offer antiperspirant protection. But their early attempts were riddled with problems, the biggest of which was caused by ACH, which had a tendency to erode the aerosol can and clog the valve when it was mixed in water inside the inner chamber.

In the early 1960s Gillette finally brought out the first aerosol deodorant. The product, known as Right Guard, was a solution-type product, similar to the war-time insecticides, and contained no trace of ACH. The product enjoyed considerable early success, and by leaving out the ACH, the company's researchers got around the can and valve issues. However, without the aluminium salt, the deodorant failed to provide any long-term prevention of perspiration. In addition, the spray was dry and often irritable to the skin; the US public preferred a moist solution.

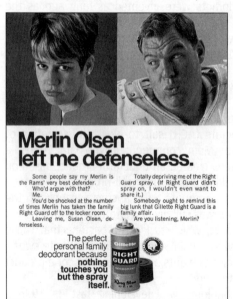

The leading cosmetics companies scrambled to find a better solution: a product that would offer genuine antiperspirant properties, as well as the anti-bacterial effect of a deodorant. Scientists searched for an alternative to water that was capable of absorbing the aluminium chlorohydrate without destroying the deodorant can and its aperture. Various aluminium

Gillette promotes its revolutionary spray technology.

compounds were liquefied in alcohol, but no combination offered the commercial viability that the cosmetics firms were after.

In the early 1960s, Gillette finally brought out the first aerosol deodorant ... known as Right Guard.

Then, in 1965, two researchers, George Spitzer and Lloyd Osipow, hit upon a solution. Their researchers suspended the aluminium chlorohydrate in oil, obviating the need for water and creating a product that would leave the can and its core components intact. Carter-Wallis heard about the Lloyd–Osipow solution, patented it and took it to market in 1967, under the name Arrid Extra Dry. The product proved popular with consumers, thanks to its antiperspirant properties and the damp, misty sensation it left on the body.

Commercial impact

Both Gillette and Carter-Wallace experienced immediate success with their aerosols. The impact of Right Guard is evidenced by the fact that in 1967, just six years after it went on the market, half the antiperspirants sold in the USA came in aerosol form. The roll-out of Arrid Extra Dry saw this figure increase again – and earned Carter-Wallace a huge share of the market.

By early 1972, aerosols accounted for 82% of all deodorant sales, and Arrid had 16% of the market, with sales of almost $65m. This early success drew new players, such as Procter & Gamble, into the market and compelled manufacturers to address a number of minor packaging flaws, with piecemeal improvements. Valve and pressure systems were tweaked to reduce bounce-off and billowing, while silicones were introduced to reduce staining.

By early 1972, aerosols accounted for 82% of all deodorant sales [but] negative publicity inflicted a huge body blow on the aerosol industry: by 1982, aerosols accounted for just 32% of the antiperspirant market

But the boom was short-lived. In 1977 America's Food and Drug Administration prohibited aluminium zirconium complexes, which had come to be used in addition to, or instead of, ACH in many aerosol deodorants. A short time later, the Environmental Protection Agency (EPA) placed severe constraints on the use of chlorofluorocarbon (CFC) propellants, another aerosol staple, due to concerns about the ozone layer. The negative publicity inflicted a huge body blow on the aerosol industry: by 1982, aerosols accounted for just 32% of the antiperspirant market.

What happened next?

The deodorant market, however, continued to go from strength to strength. In 1992, a report in the New Yorker claimed that the industry was growing by 10% year on year, and sales have remained strong as grooming assumes ever-greater importance in people's lives. According to a report released in February 2011, the UK's deodorants and body-spray market generated total sales of £548m in 2010, and market penetration was almost total. Mintel forecasts that this sector will generate sales of £581m in 2015.

During the 1980s and 1990s, it seemed that aerosols were falling away in this market as criticisms of the environmental impact of CFCs, coupled with the emergence of alternative products, began to bite. In the early 1990s stick-based deodorants accounted for 50% of the market, and it seemed that the traditional brands would be overtaken by natural, organic deodorants, which appealed to ethical consumers. When America's EPA banned aerosol products containing CFC propellants in 1993, many forecast the beginning of the end for aerosols.

However, replacement propellants have been found, such as butane, propane and isobutane, and the aerosol has clung to its foothold in the deodorant market. Some would even argue that the aerosol deodorant sector remains buoyant; a report released by Global Industry Analysts in February 2011 claimed that the spray deodorants category, which includes aerosols, constitutes the largest as well as fastest-growing product segment of deodorants in the world.

The two brands that did most to catalyse the aerosol antiperspirant deodorant have endured mixed fortunes since making their breakthroughs. In 2006 Gillette sold the Right Guard brand to Henkel in a $420m deal, and retail revenues have since remained strong; last year, Right Guard helped Henkel to increase sales by 11.2%. In contrast, Carter-Wallace, the original parent company of Arrid Extra Dry, went into a prolonged tailspin following the boom of the 1970s, and eventually sold off its various operations in 2001. Today Arrid Extra Dry is produced by Church & Dwight, and sits at the lower end of the deodorant market – a far cry from its pre-eminence 40 years ago.

17

Computer-aided design (CAD)

When: Late 1960s

Where: USA and Europe

Why: Revolutionised the way everyday machines and objects are made, through the ability to design with precision and by iteration

How: A series of innovations by academics and companies in computing

Who: Ivan Sutherland

Fact: CAD led to the technology behind the animations in the film Jurassic Park

The story of computer-aided design (CAD) involves the development of a series of innovations by a range of individual academics and engineering businesses. What they had in common was the desire for better, more efficient ways of creating new technology.

The advent of CAD software altered for ever the way things were made. It also saw the end of the time when draftsmen would draw original designs by hand for all manner of machines and objects, from cars, boats and bridges to machine parts, household appliances and microchips.

CAD designers and software can be found in engineering companies and manufacturers all over the world, and the products made using this innovation are everywhere you look.

The background

Ivan Sutherland is one of the world's most successful technology researchers, having spent a lifetime developing computing tools and innovations. It was during the early 1960s, as Sutherland was completing his PhD at Massachusetts Institute of Technology (MIT), that he invented what is now recognised as the first ever CAD software. However, Sutherland did not coin the term CAD and entitled his PhD thesis 'Sketchpad: A man–machine graphical communication system'.

The Sketchpad allowed users to design and draft for the first time on computers.

Sketchpad was remarkable in a number of different ways, although what was perhaps most revolutionary for its time was the way a user could take a light-pen and draw directly onto the screen. The system also allowed for automation in design and a memory system that facilitated the saving of master copies for duplication and further enhancements. Sketchpad showed beyond doubt that computers could be used to design and draft with superior accuracy, precision and speed to what humans could accomplish by hand.

Sutherland was far from alone in his pursuit of computer-aided tools and products. In fact, as early as 1957 another inventor, Dr Patrick J. Hanratty, had developed a system called PRONTO, a numerical tool useful in design, though it had little of the graphical sophistication of Sutherland's creation. Hanratty's ideas also involved additional quirks, such as using computer languages not commonly in use, including his own – which proved impenetrable to most.

Throughout the 1960s, aerospace companies and car manufacturers were investing in the early computers and CAD systems in a bid to make machines and products that were faster and more efficient. Universities and manufacturers were getting closer as their goals aligned, with the former providing the theory and vision, the latter providing the financial and industrial clout. In the USA much of the activity centred on the computer department at MIT. In the UK, the nucleus of the early CAD movement was at Cambridge University. Meanwhile, in France, researchers working at Renault were making breakthroughs with curve geometry.

By the 1970s, CAD technology had well and truly emerged from the R&D units and into the commercial sector, although it was still largely a 2D affair. The automotive sector was particularly enthusiastic about the benefits of CAD, and car manufacturers worked with academics to create their own bespoke software; Ford had PDGS, General Motors used CADANCE, Mercedes-Benz commissioned SYRCO and Toyota worked with TINCA. Aerospace manufacturers also had their own programs; Lockheed created CADAM, McDonnell-Douglas developed CADD and Northrop NCAD.

Throughout the 1960s, aerospace companies and car manufacturers were investing in the early computers and CAD systems in a bid to make machines and products that were faster and more efficient.

Of course, the employment of new technology was to have a considerable impact on the workforce, and people who had once worked as draftsmen soon found themselves obsolete. As the price of software decreased and the power of computers grew, so the argument for retaining people to draw by hand began to seem increasingly futile. A new breed of engineer, the 'CAD designer', began to emerge, along with a new type of company – a 'CAD company'.

CAD-users were less inclined to have their own proprietary systems designed for them, and were happy to buy off the shelf instead. Increasingly, it wasn't just the software that was in question but also the hardware on which it ran. Computers were evolving rapidly during the 1980s, and engineers were looking to develop 'the machine' that would set itself apart from the rest. Home-use of computers was also growing rapidly, and the demand from the games industry for faster, more capable machines was also driving innovation. Finally, prices were falling as competition and economies of scale lowered costs.

Computers were evolving rapidly during the 1980s, and engineers were looking to develop 'the machine' that would set itself apart from the rest.

There were winners in computing and there were, of course, losers. No company lost more spectacularly in the CAD market than Digital Equipment Corporation (DEC). At the start of the 1980s, DEC was manufacturing its range of VAX 'minicomputers', which were regularly used by CAD engineers. The company was the undisputed leader in the field, outpacing all the competition in terms of quality and price. At the time it was reckoned that DEC would be the big computer company of the decade, and its executives would have been expecting to see high profits and salaries over the next few years. But, as we now know, in an industry where technology is key, standing still is the same as moving backwards, and DEC was in for a shock.

Halfway through the decade, the industry received a heavy jolt when an engineer called Samuel Geisberg, working at the Parametric Technology Corp, created new software called Pro/Engineer for the UNIX workstation. This new 3D CAD software was a more powerful, useful and agile tool than anything that had gone before, and was much easier to use. All of a sudden, the VAX computers developed by DEC were no longer the tool of choice. Following its launch in 1987, Pro/Engineer became the number one software, and the UNIX workstation was to be the computer of choice for the industry well into the 1990s, until PCs came to dominate.

Then, in the mid-1990s, the industry shifted seismically once again. Microsoft launched Windows NT, a new form of operating system that would bring Bill Gates' company into the 21st century as the world's leading computer software company. Over the same period, Intel released increasingly powerful processors, opening up new possibilities for the tasks computers could perform. The CAD software industry now had the foundation for really exceptional development and subsequently was powered by these drivers for many years to come.

Today, CAD software is firmly embedded in our industrial infrastructure. Its development has run parallel to improvements in computing and processing power, and has also become closely tied to computer-aided manufacturing (CAM), which was actually invented during the same period.

Commercial impact

One of the most successful CAD companies to emerge during the 1970s was Auto-trol. The Texas-based business entered the CAD market in the early 1970s, and in 1979 was the first company of its type to go public. Indeed, by the end of the decade the CAD market had grown to be worth almost $1bn.

In the 1980s, as computers became more mainstream, so did CAD software. The companies involved in this industry were able to grow rapidly; companies such as Computervision, Intergraph and McDonnell-Douglas were big players, and in 1982 one of the most successful CAD businesses was started: Autodesk Inc.

The CAD software market is currently estimated to be worth in the region of $6bn, and is dominated by Autodesk, which is thought to enjoy a 36% share. There are more than five million CAD users around the world; the technology has permeated all major sectors of engineering, from cars to power stations, from ships to electronics.

The CAD software market is currently estimated to be worth in the region of #6bn.

Although most designs still begin by hand, CAD offers a faster, more precise alternative to the old-fashioned pencil as the process matures, and allows more effective communication between designers. Historically, if you wanted to change a drawing, you had to scratch out the ink, draw over it and send it back by post. But now, during the planning of a construction or engineering project, different designers can log in at various stages and update the CAD files as they go. CAD can be used throughout the design stage, from initial conceptual design to the testing of products in a virtual setting, and allows mechanical and structural engineers to work in close, smooth collaboration with architects – so any problems are ironed out before they get to site.

In recent years, CAD has begun to expand beyond the realm of engineering, into new fields of modelling and conceptualisation. Today the technology is used for everything from complex medical procedures to the manufacture of cameras, clothes and shoes, and has delivered genuine value in every industry it enters. In dentistry, for example, CAD and CAM applications are renowned for producing quick, accurate prostheses which minimise the risk of infection. Experts predict that the CAD/CAM dental market will grow to $755m in value by 2017.

CAD has also played an integral role in the rise of a number of ancillary technologies and software packages; a notable example is the product data manager, or PDM, a holistic tool which brings CAD documents together with other documents and images, allowing project managers to oversee a design programme in its entirety. According to the latest industry research, the PDM market could be worth as much as $2.8bn.

What happened next?

The growth of the CAD market is likely to continue over the coming years, as new technological advancements come on stream. Although the majority of CAD users are still working in 2D, 3D alternatives are now making rapid strides. According to a report released in August 2010, the 3D CAD market will grow at an annual rate of more than 15% between 2009 and 2013. The major factors will be the need for quicker reaction to market openings; the growing importance of collaboration between various members of the project team; the advantage of virtual prototypes and 'what-if' analyses; and a premium on high-quality design that minimises waste and the cost of subsequent re-working.

As the software continues to mature, the typical CAD application will become more proactive and intuitive. Many believe that, eventually, CAD tools will actually 'think' for the people using them – deducing what they are designing and offering pre-emptive design options as they begin each new stage of the process.

At present, CAD is generally restricted to technically savvy users, such as architects, engineers and designers. However, as the technology becomes increasingly intelligent, it may gain appeal among amateur enthusiasts, and open up a plethora of additional uses.

Many believe that, eventually, CAD tools will actually 'think' for the people using them.

18
The internet

When: 1969

Where: USA

Why: The internet has transformed global communication, making the world 'smaller' and interaction faster

How: The Pentagon's need to communicate between computers at different bases inspired the ARPANET, which transmitted the first message across a network in 1969

Who: US Department of Defense

Fact: The record for the most expensive domain name is held by Insure.com, which was bought for a reported $16m by QuinStreet in 2009

Having given rise to everything from online shopping to social networking, there's no doubt that the internet has changed our lives for the better. With hand-held devices giving us round-the-clock online access wherever we are, the internet is now so well integrated into our daily lives that it's hard to imagine life without it. However, in reality, it's only been a relatively short time since most people even gained access to the internet.

Although it was originally conceived during the 1970s, in the early stages the internet was the preserve of computer scientists and hobbyists. But that all changed in the mid-1990s, when a raft of new internet service providers made it available to a wider market. As a result, the way we communicate was radically transformed.

The background

The age of the internet is a matter for debate. Indeed, the very meaning of the word 'internet' is debatable. Some might even argue that the network of telegraph cables laid in the late 19th century constituted an early form of internet. Thankfully, if we only consider the internet as we know and use it today – a means of communication between computers – age and history are easier to pin down.

Back in 1969, the Pentagon's Defense Advanced Research Projects Agency (DARPA) was faced with a problem: it wanted to allow three of its computer terminals, at Cheyenne Mountain, the Pentagon and Strategic Air Command HQ, to communicate. The result was ARPANET, which was somewhat mythically rumoured to be able to withstand nuclear war. However, the technology it used was indeed very robust. The first two 'hosts', at the University of California and Stanford University, were connected on 29 October 1969. The first message transmitted across the network was supposed to be the word 'LOGIN' – but that was easier said than done. The first two letters went smoothly, but the system crashed at the letter 'G' – the kind of network failure that isn't entirely unheard of today.

While ARPANET continued to grow (the UK was connected to it in 1973), rival networks were springing up all over the world. In Europe, the Janet network became the preferred choice of academics, while the French developed their own version, Cyclades.

Up [until 1974], the general public were almost entirely unexposed to the internet's existence.

With dozens of new networks emerging, it was natural that someone would come up with a way to connect them all together. That person turned out to be DARPA programme manager Vint Cerf, who in 1974 proposed an 'inter-network', with no central control, that would run via a set of rules called transmission control protocol (known these days as TCP/IP) and would allow different types of machines to communicate with one another. While those rules eventually became standardised across ARPANET, it wasn't until the 1980s that the networks eventually became homogenised.

Up to this point, aside from a few articles in the press, the general public were almost entirely unexposed to the internet's existence. The problem wasn't that they weren't interested; it was more that it was still relatively complex and required a certain amount of technical knowledge. But that changed with three developments, which came in rapid succession.

First, in 1985, domain names were created. For many, that was game-changing. Although hosts were still linked to difficult-to-remember numerical IP addresses (168.192.1.1, for example), people could now reach them by typing in a word, followed by one of the approved suffixes. The first suffixes were .com, .edu, .gov, .mil, .net, .org and (naturally) .arpa. The first-ever company to register a .com address was US computer manufacturer Symbolics.com, on 15 March 1985. Some of the larger computer companies lagged behind – Apple.com wasn't registered until 1987, while Microsoft.com didn't come along until 1991. The record for the most expensive domain name is held by Insure.com, which was bought by QuinStreet for $16m in 2009. Sex.com, first registered by Match.com founder Gary Kremen in 1994, was purchased by Clover Holdings for a reported $13m in February 2011.

> *In 1985, domain names were created. For many, that was game-changing.*

The second of the three developments – and what proved to be a game-changing moment in the history of the internet – was instigated by (now Sir) Tim Berners-Lee, a scientist with the European Organization for Nuclear Research, CERN. In 1989 Berners-Lee came up with the idea for 'hypertext mark-up language' (HTML),

Sir Tim Berners-Lee, the man behind the idea for HTML and the World Wide Web, speaking at NESTA.

a programming language that allows users to 'link' between documents on a network. The project was called 'WorldWideWeb' or 'WWW', and would allow people to look at read-only information through a program called a 'browser'.

Berners-Lee decided to couple his idea with the newly developing internet, which meant that people around the globe would be able to access it. By the time the WWW was invented, there were something like 159,000 hosts connected to ARPANET, and the ability to put information on there in a series of 'pages' was revolutionary. The first web page was at info.cern.ch and simply explained what it was: 'a wide-area hypermedia information retrieval initiative'.

The final breakthrough was the creation of the first easy-to-use internet browser, Mosaic, in 1993, making the internet easily accessible to non-programmers for the first time. The browser was developed at the University of Illinois, but didn't hold the top spot for long: Netscape Navigator, which eventually evolved into a key internet security tool, came into existence the following year. By then, there were almost four million computers connected to the internet – the World Wide Web was taking steps towards becoming the service we know today.

Commercial impact

The internet's early years saw a variety of attempts to make money out of it, and the first sanctioned ads began to appear in late 1994. Examples included online magazine HotWired, which launched with ads from the likes of AT&T, Sprint, MCI and Volvo. Pathfinder, an AOL-like portal into the internet launched by Time-Warner in October 1994, also contained advertisements.

As the popularity of internet advertising grew, content publishers such as newspapers began to realise that they could monetise their online offerings by serving up ads. Soon, website ads were commonplace, with advertisers using pop-ups, pop-unders, neon-coloured slogans and even musical ditties to promote their brands. Fuelled by demand from advertisers, internet advertising became the most popular way to monetise a website. In the second quarter of 1999, online spending reached nearly $1bn.

The beginning of the dotcom bubble is generally considered to be 9 August 1995, when web company Netscape made its initial public offering, after gaining a market share of more than 90% by the mid-1990s. Having originally intended to price its stock at $14 per share, the company made a last-minute decision to up the price to $28 per share. At one stage, prices reached $75.

Hundreds of internet companies began to follow suit. In many cases, the initial public offerings were driven by entrepreneurs who were looking for rapid growth through advertising revenue, and required substantial investment from venture capitalists (VCs) to realise this strategy. One of Silicon Valley's catchphrases at the time was 'Get large, or get lost'. Indeed, it was not unknown for a business to double or even triple in size every three months.

Various experts expressed concern that entrepreneurs were ignoring the long-term prospects of their businesses. Analysts talked about the difference between the 'new' and the 'old' economies – in the new economy, profit wasn't even talked about. It was widely believed that, as long as the websites in question had the users – or the 'eyeballs', as they were referred to at the time – profit would eventually come later on in the business's life. Until then, entrepreneurs had thousands of dollars' worth of investors' money, which often went on extravagances such as flash offices. Yet the investors didn't mind, and many bought into this strategy. A common tactic was to invest in as many prospects as possible, and let the market decide which would win out.

By 2000, the NASDAQ Composite was trading at 5,048.62 – more than five times higher than anything it had traded at five years previously. No one is entirely sure what triggered the massive fall – or the 'bursting' of the dotcom 'bubble'; but on the weekend of 10 March 2000 several multi-million-dollar sell orders for major tech stocks were processed at the same time. When the NASDAQ opened on Monday morning, it had dropped by 4%, from 5,038 to 4,879 – its largest sell-off for the entire year.

Over the next year or so, thousands of companies went bankrupt as investors increasingly lost confidence in the unsustainable business models that characterised the boom – and they refused to plough in any more cash. Meanwhile, entrepreneurs struggled to rein in their costs. Online fashion retailer Boo.com famously spent $188m in just six months before going bankrupt in May 2000, while internet service provider freeinternet.com (even its name doesn't sound promising to investors) lost $19m on revenues of less than $1m in 1999, finally filing for bankruptcy in October 2000.

No one is entirely sure what triggered the massive fall – or the 'bursting' of the dotcom 'bubble'; but on the weekend of 10 March 2000 several multi-million-dollar sell orders for major tech stocks were processed at the same time. When the NASDAQ opened on Monday morning, it had dropped by 4% ...

More recently there has been renewed talk of a second dotcom bubble, with the valuations of relatively young companies with limited financial track records, such as Twitter, Facebook and Groupon, commanding huge valuations based on market-defying profit and revenue multiples. Nevertheless, the total value of sales made on the internet continues to rise, with the US Department of Commerce predicting that US e-commerce sales will exceed $188bn in 2011, whereas in the UK £58.8bn was spent online in 2010, with IMRG Capgemini predicting £69bn worth of sales for 2011.

Sites such as eBay, which generated annual revenue of $9bn in 2010, and Amazon, which turned over $34bn in 2010, account for a significant proportion of those figures. Equally, payment service provider PayPal now claims to have 100 million account holders and is forecast to reach $3.4bn in revenue in 2011.

Elsewhere, the internet has made the growth of other industries possible. Video conferencing and real-time video communication wouldn't be possible without the introduction of Voice-over Internet Protocol (VoIP). Skype, for example, reported revenue of $860m in 2010 and 145 million users a month. Media-sharing sites such as YouTube are another example. The video-sharing site streams more than a billion videos per day and is approaching $1bn in annual revenues. Online dating, where Match.com is the largest player with $343m in annual revenue, and online banking, which is said to have more than 22 million users in the UK alone, completely disrupted traditional industries. And cloud computing, which delivers online business services, is forecasted to be an $82bn market by 2016, according to research house Visiongain.

The internet has made the growth of other industries possible. Video conferencing and real-time video communication wouldn't be possible without the introduction of Voice-over Internet Protocol (VoIP).

The internet has destroyed vast numbers of traditional retailers and wiped out or heavily diluted the power of dominant players in many major consumer industries, such as music and travel – an outcome few would have believed possible just 20 years ago.

What happened next?

Today, there are estimated to be 346 million websites globally and 2.1 billion internet users around the world. Broadband made the internet a considerably more efficient experience and the demand for ever greater bandwidth remains unquenched. Furthermore, the advance of mobile devices that connect to the internet, such as BlackBerrys and iPhones, represents a massive emerging opportunity.

Online is now the primary forum for the marketing arsenal of many firms.

In 2011, in the USA alone, retailers are set to spend $220.9m on mobile devices and mobile sales volumes are predicted to reach $9bn. Furthermore, in 2010 the world's population was responsible for the sending of 6.1 trillion text messages, with automated company and public sector text messages forming a growing part of that. Such activity and the existence of relatively under-developed mobile markets prompted ABI Research to predict that by 2015 the value of goods and services purchased globally via mobile devices will hit £120bn.

ABI will doubtless have looked to emerging markets for signs of much of the anticipated growth. China, for example, reportedly has 800 million mobile web users. India is expected to have 260 million mobile internet users by the end of 2011. While in Africa, a continent where mobile phones account for 90% of all telephones in use and an estimated 500 million people have a mobile phone, smartphone penetration has much further to go.

And then there is the power of social media. The likes of Facebook, Twitter and LinkedIn have changed the way businesses communicate with their customers. These days, businesses need to be aware of how their brands are portrayed not only in the media, but also on websites such as Twitter and Facebook, where their customers are having conversations about them. Social media has become an essential PR and marketing tool for millions of firms; according to a report from analysts BIA/Kelsey, social media advertising revenues will reach $8.3bn by 2015.

Indeed, for businesses, the internet is now significantly more measurable than in its early days. Companies can now build a much clearer picture of their customers' needs and habits, making their marketing strategies a more scientific process – and in turn making advertising more targeted and relevant. Perhaps unsurprisingly, online is now the primary forum for the marketing arsenal of many firms and, far from having a competitor in its midst, the internet is growing, developing and advancing every day.

1970s

19

The pocket calculator

When: 1970

Where: USA, Japan and UK

Why: Calculations that once needed a cumbersome computer could be performed on the go, greatly increasing efficiency

How: Advances in mathematics, micro-processing and other fields

Who: Texas Instruments

Fact: Between 1971 and 1976 the cost of a pocket calculator fell by over 90%

I n its heyday, the pocket calculator became a must-have for every schoolchild, office worker and professional across the Western world. It was a symbol of how technological innovation was impacting on everyday life. Today, although still present, the hand-held tool has been replaced, for many people, with the software found on PCs, laptops and mobile phones.

The 'single use' calculator now seems like overkill in the age of multi-function devices (MFDs) and the internet. However, these early forms of computing were a democratic force, bringing processing power to the fingertips of the masses.

The background

Throughout the 1960s, inventors and manufacturers worked intensely on developing the calculator. The portability, power supply, functionality and accuracy of the device underwent considerable change and development and the early predecessors of the pocket calculators came to market. However, these gadgets, though ground breaking, were a far cry from those that would be made in the following decade.

The 1970s saw a ferocious struggle for supremacy between calculator manufacturers, [which] took place in R&D labs and on the high street.

Most 1960s calculators were desktop mounted (too heavy and too large to be fully portable), required a mains power supply and were provided with paper print-outs rather than electronic displays. They were also too expensive to be truly mass-market products and were largely bought by businesses and professionals rather than consumers. However, by 1970 most of the research and development required to make a pocket calculator for the masses had been achieved. Much of this work had been carried out by researchers at Texas Instruments (TI), most notably by Jack Kilby, who created integrated circuitry.

The 1970s saw a ferocious struggle for supremacy between calculator manufacturers, and this struggle took place in R&D labs and on the high street. There would be many winners and losers as manufacturers pursued consumer demands and requirements. Much of what was achieved in the pocket calculator market would be seen again almost 30 years later in the mobile phone arena; manufacturers strove to create cheaper and smaller devices, while at the same time trying to pin-point customer requirements for functionality and also aiming to solve the issue of power supply and battery life.

Commercial impact

In late 1970 and early 1971, the first pocket calculators came onto the market. But whether these really were 'pocket' calculators would depend on the size of the consumer's pockets, both physically (they were large and heavy) and metaphorically (each cost a few hundred dollars). One of the most notable was Canon's Pocketronic, which used chips developed by TI. Also coming onto the market in 1970 was the Sharp Compet QT-8B, recognised as the world's first battery-powered electronic calculator.

TI, which had led the way for the industry with the development of its 'Cal Tech' experimental calculator, did not enter the consumer market until 1972. However, when it did, it did so in style, with the TI 2500 'Datamath', followed closely by the TI 3000 and the TI 3500. These early models came with LED displays and rechargeable batteries, weighed less than a pound and retailed at less than $150 – truly a mass-market device. It is perhaps only fair that TI's pioneering work in the late 1950s and 1960s led to its gaining the patent for the hand-held calculator in 1974. However, the battle for market share would continue, as would the race to reach key technological milestones.

During the early and mid-1970s, companies from all over the world had entered into a booming pocket calculator market. Casio, Commodore, Rapid Data, Lloyds, Digitrex and HP all began to compete with TI, Canon and Sharp. One of the main British companies to push the market forward was Sinclair, led by the engineering genius Clive Sinclair. The Sinclair Executive was the first truly slim-line calculator (it could fit into a pocket) and was also affordable for the masses.

Utilised by Sharp Electronics in Japan, LCD displays (liquid crystal displays) were the pocket calculator's next innovation, requiring vastly less power to operate and hence improving battery life. This, in turn, meant that additional functions could be added beyond the four standard uses (addition, subtraction, division and multiplication). Increased competition and the decreased cost of components were also pushing the

The Sharp QT-8B 'micro Compet,' one of the first hand-held, battery-powered calculators.

price down and, towards the end of the decade, units could be bought for 20% of their 1972 price.

In 1975, New Scientist claimed the pocket calculator market was worth $2.5bn. However, during the turbulent years of the 1970s, making predictions on market value was a dangerous game. Furthermore, price-cutting and competition meant that many companies abandoned the market, such as Sinclair, which enjoyed success in the more lucrative personal computer market. Manufacturers also overestimated market demand and there were several reported cases of warehouses being full of unwanted products.

What happened next?

By the end of the 1970s, the rapid development of the pocket calculator was largely completed, including the development of solar cells, which overcame the problems of battery life. The market had undergone a disruptive period, followed by consolidation, and the big players would continue to compete throughout the 1980s and beyond. The major players TI, Casio, Canon and HP are still active in the market but their interests have now moved beyond pocket calculators.

Pocket calculators have remained with us to this day because they are easily affordable, inherently useful, and so user-friendly that anyone (including young children) can use one without tuition.

Electronic goods are now a huge part of our lives, both at work and in our leisure time. Many of the companies that came to prominence through devices such as the pocket calculator have long since moved into other, more lucrative areas, such as home computing.

Pocket calculators have remained with us to this day because they are easily affordable, inherently useful, and so user-friendly that anyone (including young children) can use one without tuition. However, to this day some educators complain that their introduction into the classroom has meant that a generation has grown up thinking that basic numeracy is unimportant. Proponents, meanwhile, argue that in fact they encourage an appreciation of maths. But what is clear is that calculators will remain with us in one form or another for as long as we need to do sums – and there is little prospect of that ending any time soon.

20
Budget airlines

When: 1971

Where: USA and Europe

Why: Budget airlines transformed the airline industry, substantially growing passenger volumes

How: Entrepreneurs overcame the old, vested interests of government and its flag-bearing carriers. Where Southwest Airlines' Herb Kelleher led, others followed

Who: Southwest Airlines

Fact: The aerospace industry accounts for 8% of global GDP

Budget airlines seem to be loved and hated in equal measure. Loved by many consumers because they offer less-affluent travellers a cheap and fast route to destinations that were previously out of their reach. But hated by environmentalists and customers who dislike the poor service delivered by some, or who didn't appreciate the small print and got stung by hidden charges or extras.

Budget airlines have a reputation as a rebellious, sometimes cocky bunch that seem to be in near-constant battles with regulators, competitors and whomever they have rubbed up the wrong way this week. But the no-frills sector is now a considerable, profitable and growing part of air travel and looks set to remain so for some time to come.

The background

Domestic air travel began in earnest after World War II and jet engines for civil use were being deployed by the 1950s. Throughout the 1960s and into the 1970s, the majority of airline companies were 'flag carriers', often sponsored or owned by their respective governments. Some of them retained a rather patriotic, if not regal, disposition. Pilots were often former military flyers, and this only added to the sense that airlines were part of the establishment.

For many, British Airways epitomised this exclusive approach to air travel. Those who travelled by air during the early years of the industry were generally at the wealthy end of the scale, and the actual flight was seen as an outing in itself; passengers would dress smartly for the occasion and would expect a meal and a drink as part of the experience. High fares were generally accepted as an inevitable consequence of this luxury method of travel. But it was not long before entrepreneurs outside the industry began to see potential gaps in the market for a new type of service.

Coinciding with the increased sense of emancipation prevalent in the late 1960s was the growth of the 'package holiday', where tour operators would charter entire planes to fly their customers as part of an all-in-one deal. One of the first such operators was Euravia, which began flights from Manchester in 1961. Other tour operators, such as Thomas Cook, soon followed. The industry really took off in the early 1970s, and with it, tourism to previously seldom-visited destinations such as Crete and the Algarve.

Coinciding with the increased sense of emancipation prevalent in the late 1960s was the growth of the 'package holiday'.

But it was in the USA that a true revolution unfolded in airlines: a model that emphasised lower cost over luxury by cutting down on extras, using smaller airports to save costs and abolishing seat booking. The world saw the rise of the budget airline, starting at a small airfield in Texas, a phenomenon that was to change the way passengers viewed travel and that would open up aviation to the masses for the first time.

Southwest Airlines was the world's first no-frills budget airline, and pioneered the business model that brought it and others that followed so much success.

Southwest Airlines flight attendants sported hot pants in the 70s.

It was established in 1967 by a group of investors headed by Herb Kelleher, and planned to operate business flights within the state of Texas, between Houston, Dallas and San Antonio. However, because of protracted legal challenges by the state's major airline companies, Southwest did not actually open for business until 1971. Initially offering flights between Dallas and Houston, the airline made the decision to drive down fares by stripping away the extraneous luxuries that beforehand had been considered part and parcel of flying. There was no free meal or drink, the airline launched with a single passenger class and there were no reserved seats, meaning it could fill more planes and drive down costs. It also flew from 'secondary' airports rather than the more expensive, higher-profile main airports, further cutting costs. Initially, a one-way ticket cost just $20 and the airline offered flights on the hour between Dallas and Houston, making the service both cheap and convenient for business travellers.

And the company was keen not to present itself as a staid, puritanical airline, which would seemingly have been in keeping with a 'no-frills' service – indeed it was quite the opposite. In reference to its home base, Love Field in Dallas, the company began cultivating a corporate image of 'love' – stewardesses were hand-picked for their perceived attractiveness and were attired in hot pants and go-go boots, appealing to Southwest's client base of mostly male business people (this practice stopped in 1980, when the stewardesses won the right not to wear hot pants on the job). Southwest also decked its planes out in recognisable red, blue and orange paint, providing instant brand recognition even when thousands of feet in the air – a practice followed by many budget airlines today, such as easyJet, with its distinctive orange livery.

Soon, other airlines in Texas caught on to the budget model and in 1983 Southwest became embroiled in a price war with rival Texas operator Braniff. This resulted in customers seeing return fares from Houston to Dallas drop to as little as $13, as well as being offered extras like free liquor and ice buckets in return for their custom. The budget airline was well on its way to being big business.

When Congress deregulated US airlines in 1978, Southwest became free to expand its operations outside of the state, and immediately set about purchasing other domestic flight carriers to extend its reach across the USA, starting with the purchase of Midway Airlines in Chicago in 1978.

The first no-frills airline to cross the Atlantic was established by a British entrepreneur called Freddie Laker (later Sir Freddie). Laker Airways was set up in 1966, initially as a charter airline. However, Laker conceived of a fast and stripped-down service that would fly from London to New York on a daily basis. Seats would be sold on a 'first come, first served' basis because there would be no booking, and tickets would be just £30. Laker called his plane the 'Sky Train', and applied for a licence to run it in 1971. However, the British authorities just weren't ready for Laker's plan and constantly put up barriers and restrictions on his business. People were concerned about the number of flights taking off from Heathrow because of the safety issues involved, as well as the pollution and noise they could cause.

The first no-frills airline to cross the Atlantic was established by a British entrepreneur called Freddie Laker ... Laker called his plane the 'Sky Train'.

The Sky Train did eventually get off the ground, although it did not fly as often as Laker wanted and had to take off from Stansted rather than Heathrow. Other transatlantic operators also played rough by dropping prices in an attempt to put Laker out of business. Throughout its duration, Laker Airlines was caught up in disputes with the Civil Aviation Authority (CAA) and government. Laker's vision was not compatible with the uncompetitive, closed-market approach of the day. In 1982, after a mixed history, Laker Airways went bust, although its founder had won the affection of many as a result of his efforts.

Commercial Impact

The main carriers might have seen Laker go out of business, but new entrants were lining up for a stake in the growing sector. At the start of the 1980s, the air travel market was still dominated by a few big players and almost half of all flights took place within the USA. However, a number of crucial changes paved the way for the booming industry we have today. These changes had broader commercial implications for the tourist industry.

For one thing, legislative changes, such as deregulation of the airline industry in the late 1970s and the privatisation of many formerly state-owned carriers, such as British Airways, into the 1980s made it far easier for new entrants to get into the sector and challenge the existing order. In the past, governments had controlled factors such as the numbers of flights, air fares and destinations, but in the USA, Europe, Australia, Japan and parts of South America, these restrictions gave way to market forces.

Secondly, business cycles finally began to work in the favour of new entrants. The recessions of the late 1970s, the early 1980s and again in the early 1990s meant that start-ups could obtain aircraft hangar space at a much cheaper price and could also find experienced staff to work for them who had been victims of the economic slowdown.

The political landscape of Europe changed dramatically with the collapse of

Communism in Eastern Europe in the early 1990s. This opened up a host of new destinations to travellers and expanded the need for new airports and planes. Further business deregulation, especially the Single Market, created by the European Economic Community (EEC) in 1992, added further to the sense of freedom and openness, and destinations such as Prague – previously virtually inaccessible – experienced a huge and sustained boost in tourism.

During the 1980s and 1990s, the cheap flights and no-frills business models were picked up by many providers across the world. One of the most successful was Ryanair, founded in Dublin in 1985. The company began by offering flights from Waterford, Ireland to London. Ryanair soon started to grow, but was not particularly profitable. However, this changed in 1988 with the arrival of CEO Michael O'Leary. O'Leary had learnt of the low-cost model adopted by Southwest Airlines and was convinced that it could work for Ryanair. He was proved right and, by 1997, was ready to take the company public and to use the money to make a major challenge to the airline industry. By 2003, revenues were in excess of £1bn and the company was highly profitable.

The 1990s also saw the growth of Stelios Haji-Ioannou's easyJet airline. Founded in 1995 and initially offering domestic services to London and Scotland, it won over many passengers with its commitment to making air travel 'the cost of a pair of jeans' and gained exposure in the early years by simply displaying a huge phone number on the side of its planes. In 2003, it acquired rival budget operator Go Travel for £374m, almost doubling the number of planes in its fleet. easyJet grew to carry more passengers than any other UK-based airline and the distinctive orange planes can now be spotted at some 118 airports around the globe.

Nowadays, Southwest is the USA's biggest airline and runs 3,400 flights a day to destinations all over the country, reporting revenues of $12bn in 2010.

What happened next?

Over the last 60 years, air travel has grown and grown. It is now a major part of the world economy, directly employing over five million people. There are currently roughly 900 airlines in operation, and between them they have approximately 22,000 planes. It is hard to measure the impact of air transport on the global economy, but groups such as the Air Transport Action Group believe it to be 8% of the world's GDP, or $2.9bn. Budget airlines have played a massive part in this growth and have, themselves, become major companies. Ryanair, now Europe's biggest budget airline, carried 33 million passengers during 2008 and currently has revenues of about £360m per year.

21
Email

When: 1971

Where: USA

Why: By providing a cheaper, quicker and more convenient form of correspondence, email has changed the way we communicate for ever

How: Scientists and technologists wanted to create a method of exchanging digital messages electronically

Who: Many people are credited with contributing to email's evolution, but Ray Tomlinson is often hailed as its original inventor

Fact: Queen Elizabeth II was the first head of state to send a message by email

For most people, it's hard to imagine life without email. Whether it's to communicate with colleagues or clients or to make plans with friends for the weekend, we all rely heavily on this digital method of communication. You only need to see the pained expressions on colleagues' faces when the email server goes down in the office, to grasp its significance to modern living. Globalisation and the increased popularity of outsourcing mean that colleagues are often scattered across the globe: email is an instant way of communicating with them without the cost, or hassle, of organising face-to-face meetings or conference calls.

The background

Although, for the most part, email (short for electronic mail) has only been part of our everyday lives for the last couple of decades, it's actually been knocking around, in one form or another, since the early 1960s. Not only does it pre-date the internet as we know it today; it was arguably a key tool in its creation.

Electronic message exchange in its most basic form has existed from the old days of timesharing computers, which were first unveiled at Massachusetts Institute of Technology in the USA in 1961. This technology enabled multiple users to connect to the same mainframe computer via remote dial-up. In 1965, email was devised to enable the various users of a time-sharing mainframe to communicate with one another through a process that was as simple as leaving a memo on a colleague's desk.

This early form of email enabled hundreds of co-workers to correspond. But the messages could only be exchanged between people working on the same mainframe, and there was little or no interoperability. Even when the technology developed as far as allowing inter-organisation messaging, mainframes could only communicate if they ran the same email system and proprietary protocol. One of the biggest glitches was that, for computers to communicate with each other across different networks and systems, they needed a way to identify each other. Just as it is necessary to write an address on an envelope before a letter is put in the post-box, computers needed to identify where they should be sending these important, and sometimes sensitive, communications.

In 1971 Ray Tomlinson, an engineer working on the development of ARPANET, the antecedent of today's internet, set to work on a system for sending emails between separate computers. In a process that he claims took only a few hours, Tomlinson took a program called SNDMSG, capable of sending emails over a local system, and merged it with another program called CPYNET, which had been developed to send files to other computers connected on ARPANET.

Tomlinson then picked the '@' symbol from a keyboard to separate the name of the user or recipient of the email, and created the simple convention 'user's name-@-their computer's name' for email addresses. In October 1971 he sent the first-ever email from one machine to another, sitting right alongside it, using the ARPANET connection. Although Tomlinson's memories of the event are vague – he doesn't even remember the content of the message – he is now widely credited as the 'father of email'. And through the various incarnations of email and the internet, the '@' sign has endured, becoming a symbol of the digital age.

Electronic message exchange in its most basic form has existed from the old days of timesharing computers, which were first unveiled at Massachusetts Institute of Technology in the USA in 1961.

Email quickly became the most popular application on the ARPANET, and began to generate interest from further afield. Some of the early adopters of email were in the US military because it was a good way to communicate with comrades stationed at other bases. And it wasn't long before its use spread to the corporate world, with the emergence of tailored commercial packages. By 1978, 75% of ARPANET traffic was email.

Another key technological, and social, development that influenced the evolution of email was the rising popularity of the personal computer, or PC. These computers contained a piece of software called an offline reader, which allowed the user to store their email on their own machines – where they could read them and prepare replies – without actually being connected to the network, almost like an early version of Microsoft Outlook. This was particularly useful in places where connection costs (basically the price of a telephone call to the nearest email system) were expensive; once an email had been written, users only needed to connect in order to send it.

In the early 1980s a flurry of new interconnected computer networks, inspired by ARPANET, began to spring up, creating the need for a set of industry standards that were capable of fostering communication between the new networks. In 1982 the simple message transfer protocol, or SMTP, was established as a standard for the sending and receiving of email; two years later, the Post Office Protocol, or POP, was created to homogenise the storage of email. Both formula were fairly basic – in fact, SMTP provided no way of showing whether a message had been received or read, and made no attempt to find out whether the person sending the email was who they claimed to be. But these early standards played a crucial role in building bridges between the various networks; anyone could now send email to anyone else, simply because they used the same essential protocol.

With common standards in place, people began to wake up to the potential of email; it was clear that this new form of communication could spread far beyond the academics and computer enthusiasts who comprised the bulk of its early user base. In 1983 MCI Mail, widely credited as the world's first commercial email service, opened for business with a grand press

The Lapel button was used to market the MCI Mail electronic mail service, ca. 1985.

conference in Washington, DC. Other services followed suit; in the late 1980s two of the most recognisable early email systems, Pegasus and Microsoft Mail, went live to the public, while CompuServe developed an email service for its subscribers. Then, in 1992, Microsoft Outlook, an integral tool for business email, went live.

These easy-to-use, inexpensive services played a crucial role in the subsequent propagation of email. Whereas in the past a specialist service was needed to send emails, a host of companies were now giving away email accounts and inviting users to send and receive messages, with no need for technical knowledge or pricey equipment.

Much faster and cheaper than conventional mail, and with the ability to send attachments such as pictures, documents and spreadsheets, the advantages of email rapidly struck businesses around the world. Sending out copies to multiple recipients was considerably easier with email than with any other form of communication.

With global internet usage increasing to 16 million people in 1995, and then to 147 million people in 1998, all the conditions were in place for email to become a cornerstone of people's daily life. In 1990, around 15 million email inboxes were in use worldwide; by the end of 2000, this had increased to 891 million.

Commercial impact

As more and more people began to connect to the internet in the mid-1990s, email quickly became big business. In 1997, just a year after it launched, Hotmail was sold to Microsoft to $400m – demonstrating that the market leaders were willing to pay vast sums for a simple, accessible product tailored to the email market.

The creation of more sophisticated anti-spam tools did little to check the rise of email marketing, which has grown steadily, in line with overall email use. The roots of email marketing were laid in the 1980s, when bulletin boards, the precursors of modern online forums, began to receive messages about promotional offers and product launches. By 1995, the number of advertisements sent by email was outstripping the number sent out by regular mail, and the volume of marketing messages has continued to increase exponentially over subsequent years.

A 2010 study from technology market research house Radicati Group suggests that 294 billion emails are sent every day.

But of course email's real significance for business was much broader than that. The wider commercial impact on business is impossible to calculate; there's no way of adding up the number of sales, mergers and staff hires that have been triggered by email correspondence. However, considering that around 730 million corporate email accounts were in use in 2010, and the average employee sends more than 30 emails per day, it's safe to say that email has literally transformed the way people do business throughout the world.

Email systems such as Microsoft Outlook have evolved into fully fledged personal organisers, placing email at the heart of people's daily lives, and the emergence of chat-based programmes such as Windows Live Messenger, not to mention social media, video-conferencing and online dating sites, has made online correspondence a familiar convention for millions.

A 2010 study from technology market research house Radicati Group suggests that 294 billion emails are sent every day; research released in late 2010 found that one in five British workers spend 32 working days a year managing their email; it is believed that the typical corporate user sends and receives around 110 email messages a day; and, according to some experts, a billion new inboxes will be opened between now and 2015. Every aspect of email is now extremely lucrative – the global IT security market, which focuses heavily on email, is now worth $16bn, while the value of email marketing has reached £350m in the UK alone.

The rise of email has also led to the demise of 'snail mail'. Around the world the volume of conventional letters posted has fallen steadily, threatening the viability of postal services in many mature economies.

What happened next?

Most of the business world relies upon communications between people who are not physically in the same building or city or even on the same continent – especially in the 21st century, when globalisation means that teams are more likely than ever to be geographically dispersed and businesses of all sizes require cheap, flexible correspondence tools.

Furthermore, the pace of change shows no sign of slowing down. Perhaps the biggest single change in recent years has been the emergence of the smartphone, which enables users to check their emails on their phone's screen. The convenience of the smartphone, which allows people to browse their messages from almost anywhere on earth, has led to huge market take-up; indeed, a study in May 2011 found that mobile email use had increased by 81% in less than a year.

The development of cloud computing, which allows companies and individuals to access key applications online and avoids the need for expensive desktop software, is another key change. Cloud computing allows a business to 'rent' an email server, paying an external party to manage their email system and install all the necessary software updates. Microsoft claims that a business employing 1,000 employees could save $2m a year through cloud-based email; this is sure to persuade thousands of firms to sign up over the next few years.

The convenience of the smartphone, which allows people to browse their messages from almost anywhere on earth, has led to huge market take-up.

What will be the next seismic change? Some technology forecasters predict the merger of conventional email into social media platforms, such as Facebook or Twitter, while others dismiss this. Social networking giant Facebook recently launched its new 'social inbox' feature. This new unified messaging system funnels both the online and mobile communications methods people use – email, SMS, instant messaging, Facebook chat messages – into a single place.

Facebook isn't the first to try something like this: web browser Mozilla has been experimenting with it for a while. And there's a reason no one's quite got it right yet – having a single interface capable of interacting with a variety of different messaging mediums is tricky, and expensive, to get right. Time will tell if Facebook's social inbox will prove the hit Mark Zuckerberg hopes it will be, though the jury remains firmly out at the time of writing. Either way, email has become a significant part of business life, and seems set to remain important, however it develops next.

22

The automated teller machine (ATM)

When: 1971

Where: USA

Why: The invention changed how the world accessed cash

How: The inspiration came to Docutel product planner Don Wetzer while he was waiting in a queue at a Dallas bank

Who: Don Wetzel, Tom Barnes and George Chastain

Fact: The number of ATMs currently in use worldwide is estimated at over 1.8 million

Cashpoints, holes-in-the-wall, call them what you will: ATMs (automated teller machines) are that modern convenience it's hard to do without. Thanks to a simple smartcard and PIN, it's perfectly possible to go about your daily business without ever darkening the door of a bank. Access to cash at any time, whether paying for client drinks, shopping in the afternoon, or partying at 3am: it's so obvious an idea it's hard to believe it only rolled out in 1971.

The nondescript ATM was a business idea that, quite simply, changed the way the world accessed cash. It made banking a 24-hour operation and marked the beginning of the electronic banking revolution.

The background

Before the advent of the ATM, it was necessary for customers to visit a bank or financial institution to access the money in their accounts. Since tellers only worked business hours and banks often closed on weekends, there was simply no way for customers to access their bank accounts during these times; and it was not unusual to find long queues outside the major banks come Monday morning.

The ATM was not the brainchild of one inventor alone. While America's Don Wetzel and company Docutel are credited with the introduction of the first modern ATM, many bright sparks around the world were making efforts simultaneously, fighting it out for the cash-machine crown. Way back in 1939, one Luther George Simjian, a Turkish American, registered a whole series of patents relating to a cash-dispensing machine. A prototype Bankograph, a predecessor of the ATM that enabled the paying of utility bills, was installed in New York banks in 1960, and in 1965 CHRON developed an automated vending machine capable of accepting credit cards. Then, in the late 1960s, Scottish inventor John Shepherd-Barron introduced the first operational cash machine in the UK.

It took a while for the idea of ATMs to really take off. At first, banks were even deterred by the high cost of the machines in comparison to human bank clerks.

A forerunner of the ATM that we know today, Shepherd-Barron's inaugural De La Rue Automatic Cash System appeared in 1967 at a north London Barclays Bank. Calling late to the bank on a Saturday afternoon, Shepherd-Barron was unable to access his funds: this was the germ that inspired his voucher-based cash machine. Apparently, a tech-savvy Barclays' manager was convinced to trial the prototype 'over a pink gin'. The De La Rue predated the machines we are now familiar with, using chemically-treated cheques and PINs rather than the plastic card and magnetised strip that we know today. The machine also had no direct connection to the bank's security and computer systems. It was an American who took the cash machine to the next level.

Kazuma Tateisi, the founder of Omron with an early ATM dubbed the 'Money Machine'.

The Smithsonian's National Museum of American History credits the Docutel Corporation, the company behind automated baggage-handling equipment, for the first networked ATM. This machine, conceptualised by Don Wetzel, vice-president of product planning at Docutel, and devised by chief mechanical engineer Tom Barnes and electrical engineer George Chastain, took some $5m to develop. It required customers to use a plastic payment card, such as Mastercard or Visa, to access their money. This magnetically encoded plastic, since superseded by a smart chip, carried customer information, and reduced the risk of fraud. The prototype was rolled out in 1969 and the 'Total Teller', the first fully functional bank ATM, was introduced by Docutel in 1971. The UK quickly followed suit, with ATMs first commissioned by Lloyds Bank coming into general use late in 1972.

Surprisingly, considering their convenience, it took a while for the idea of ATMs to really take off. At first, banks were even deterred by the high cost of the machines in comparison to human bank clerks. Incredibly, there were no more than around 5,000 machines in use in America by 1975. Only in the 1980s, with the establishment of interbank networks enabling the instant transfer of funds between banks, did the cash machine really catch on.

Commercial impact

From the 1980s on, ATMs and their manufacture became big business, and the industry is currently dominated by American companies, from IBM to NCR. The

One of BT's converted cash machine telephone booths.

ATM market is global and continues to grow across various regions, from Asia/Pacific to Europe and Latin America. But it is the impact of the ATM, rather than the machine itself, that is the big idea business-wise.

ATMs were the first big step towards large-scale automation in the services industry. The success of the cash machines signalled to businesses everywhere just how much more efficient it was to have customers serve themselves. With the introduction of ATMs, the requirement for bank tellers to work business hours was suddenly eliminated, and the way was cleared for other services to follow suit. Nowadays, one can pay bills, take out small loans and transfer credit, all at a street-level ATM kiosk.

A recent trial in the UK has even found BT converting disused telephone booths into cash machines. And the idea of using one technology to house another is increasingly being mirrored by the ATM itself. The cashpoint has moved on from being a simple paper money dispenser to becoming a truly multi-faceted machine, with a broad range of functions. These days, partnerships are being forged between financial institutions and mobile phone operators, lottery providers and postal companies. On a visit to the cashpoint, customers can use ATMs to purchase mobile phone credit, to donate to charities and even to buy train and lottery tickets. One cash machine in London even dispenses gold bullion. The ATM is increasingly being used by advertisers as well, displaying marketing messages on the terminals.

The success of the cash machines signalled to businesses everywhere just how much more efficient it was to have customers serve themselves.

Another interesting feature of cash machines is their use by businesses as sources of in-house revenue streams. While a great number of cash machines are deployed by financial institutions, and are on the same site as the bank, many other ATMs are off premises. A huge range of companies, from corner shops to pubs and restaurants, install cashpoints in their premises in order to attract new custom and increase existing business. Research indicates that the typical ATM customer may spend up to 25% more than a non-ATM customer, but this is not the primary draw. Many ATMs provide additional money-spinners for small businesses, through surcharges, on-screen advertising and related dispensing services. The ATM is not just a cash dispenser, it's a serious money maker.

The electronic transaction system pioneered by the ATM showed the business world the potential of financial connectivity in related fields, and proved to be the catalyst for other innovations, such as Electronic Funds Transfer, a method of automatically transferring funds from one account to another, which led to telephone banking and e-commerce.

What happened next?

For now, at least, there remains a need for cash, and the ATM remains a fixture on any high street. But in the coming years, its ubiquity may come to an end. The ATM is over 40 years old, and taking into account the explosive pace of developments in modern technology, that can almost be deemed ancient. The inventor of the De La Rue automated cash machine, John Shepherd-Barron, died in May 2010: during his lifetime, he saw the rise and rise of his brainchild, and that of his rival Wetzel and Docutel.

The ATM is presently such a fixture of the streetscape that it is hard to imagine life without it.

Despite its multi-faceted service options, its adoption of third-party services and advertising possibilities – not to mention its real ubiquity – the ATM is showing its age. Most large retailers and increasing numbers of small ones employ the use of card readers in their stores, meaning that cash is increasingly becoming unnecessary – especially in the light of continuing innovations in cashless payment, such as contactless debit cards, mobile payment systems and online banking. Already in Japan, there are over six different cashless payment systems.

The ATM is presently such a fixture of the streetscape that it is hard to imagine life without it. But given time, this symbol of automation may itself become as rarely used as the iconic telephone box now used to house it.

23

The computer game

When: 1971

Where: USA

Why: It created an entirely new market that is now worth billions of dollars

How: Computer Space was the first commercially sold coin-operated video game and was released by little-known firm Nutting Associates. The game's creators founded Atari, Inc a year later

Who: Nolan Bushnell and Ted Dabney

Fact: The average gamer spends 18 hours a week playing video games omputers were originally developed to automate repetitive administrative tasks. But the people who built and wrote programs for them quickly realised that they were also extremely good for enhancing their leisure-time pursuits. Initially played almost exclusively by boys and young men, games have grown steadily and spectacularly into a mass-market leisure activity now enjoyed by men and women of all ages. Such is its position now that the video gaming industry has revenue greater than the film industry, and around 1.2 billion now play games around the world.

The background

The computer games industry truly began in 1971 with the debut of the first commercially available game, though there were some rudimentary games available for main-frame computers before then. The world's first commercial computer game, a space combat simulator called Computer Space, was produced by Mountain View, California-based Nutting Associates, a producer of mechanical coin-operated games for arcades. The game enjoyed only very modest success and its creators, Nolan Bushnell and Ted Dabney, left Nutting Associates after falling out with its owner over equity in the company.

Atari founders (from left to right) Ted Dabney and Nolan Bushnell with Larry Emmons and Al Alcorn.

Convinced of the commercial potential of this new entertainment medium, Bushnell and Dabney founded their own company – Atari, Inc in 1972. Atari's first game was actually a test for a new member of staff Bushnell and Dabney had hired. They liked it so much that they decided to launch it as their first product – the now legendary coin-operated arcade game Pong. This tennis simulator was simple to understand and play – the only controls in the original arcade version are two knobs to move the paddle. It quickly captured the public's imagination. Regulars at the Silicon Valley bar where the prototype machine was installed queued outside to play and quickly broke it, filling it with so many coins that it shorted out. Atari went on to sell more than 19,000 Pong cabinets and spawned numerous competitors, giving birth to the modern computer games industry.

The world's first commercial computer game, a space combat simulator called Computer Space, was produced by ... Nutting Associates, a producer of mechanical coin-operated games for arcades.

Atari grew rapidly, and very profitably. Bushnell bought out Dabney's share of Atari and Dabney returned to repairing pinball machines, his role before video games. In 1977 Atari wanted to launch a small console that people could plug into their television sets to play video games in their homes. Needing more capital to launch this effectively, Atari was sold to Warner Communications. The console, named the Atari Video Computer System, was launched in 1977 and became so successful that at its peak Atari was the fastest-growing corporation in the USA.

Meanwhile Bushnell and Dabney's previous employer had hit hard times. The visionary Nutting Associates went the way so many first movers do – out of business.

One of the new companies set up to compete with Atari after Pong's early success was Taito, formed in Japan in 1973 by Russian immigrant Michael Kogan. In 1978 it launched a ground-breaking game designed by Toshihero Nishikado. Space Invaders tasked players with defending the Earth against endless waves of aliens, armed with a laser cannon. The game was an instant hit in its native Japan – according to popular legend, the Japanese government was forced to mint more 100-yen coins to cope with a shortage caused by so many people playing the game. By 1980, there were 300,000 machines in the country and a further 60,000 in the USA.

If some of Pong's success can be explained by its two-player mode, pitting one human player against another, one of the reasons why Space Invaders became a global phenomenon – quite apart from its exciting and innovative gameplay – was its high-score screen, compelling players to keep playing to top the leader board.

Commercial impact

Keep playing they did. Space Invaders generated an estimated $500m in profit for Taito by 2007, making it the most successful video arcade game of all time. It also ushered in a golden age of gaming – thanks to the popularity of games such as Asteroids, Frogger, Donkey Kong and Pac Man (one of the first games with cross-gender appeal, bringing females to the previously male-dominated arcades), the industry generated $11.8bn in North America in 1982, more than the annual revenue of its film and pop music sectors combined.

Video games grew rapidly alongside the expanding home computing market. In the 1980s, gaming moved out of the arcades and into the home, thanks to the popularity of the early consoles such as the Atari 2600 and its progeny such as the Nintendo Entertainment System, and the Sega Master System, as well as home computers such as early Tandy and Apple machines and Commodore's Vic 20 and Sinclair's ZX Spectrum computers. By 1984, home gaming had surpassed the arcade in revenues.

This era was a golden age of video games, spawning not only many classics but also new genres.

The massive growth in home computing turned what had been a hobby and a cottage industry in the 1970s into a fully fledged business in the 1980s, with numerous dedicated software and hardware companies springing up to provide new games and accessories for this burgeoning market. Electronic Arts, one of the world's largest and most successful computer game software publishers, was set up in 1982. This era was a golden age of video games, spawning not only many classics but also new genres. The Legend of Zelda helped to define the action-adventure; Kung-Fu Master the side-scrolling beat 'em up; Elite the space flight simulator/trading game set in a huge universe; and Rogue, Akalabeth and Ultima, the role-playing game. Many of these games involve puzzle solving and are entirely peaceful, in stark contrast to the much-perpetuated image that all computer games are violent. These new genres broadened the appeal of the sector to many more people, fuelling further growth.

What happened next?

The technology enabling video games developed incredibly fast, advancing notably every year. This led to a series of booms and busts throughout the

1980s, 1990s and 2000s, frequently resulting in different manufacturers leading the market with every new generation of console or computer. Home-computing pioneer Commodore went out of business, while Atari eventually ceased selling consoles or computers and was eventually sold for just $5m for its brand names. Nintendo emerged as a market leader by introducing a different sort of controller instead of the traditional joystick, and by selling its consoles very cheaply and instead making its money from the games the new console owners went on to buy. Sega, initially a major arcade game producer and briefly a leading console manufacturer, now focuses on developing games, leaving Nintendo and relatively recent entrants Sony and Microsoft to compete for new console sales.

> *Nintendo emerged as a market leader ... by selling its consoles very cheaply and instead making its money from the games the new console owners went on to buy.*

Developing software has proved similarly risky, with the majority of games losing money and publishers relying on earning more from their few hits than they lose on the misses. Typically, as with hardware manufacturers, the leading software developers for one generation of computing technology have hit financial difficulty when a new generation of hardware launches, giving rise to new market leaders. Electronic Arts is almost unique in having managed to stay at the top almost ever since its launch; its revenues in fiscal year 2011 were $3.6bn. Today, top games for PCs or consoles cost tens of millions of dollars to develop and sell millions of copies each, generating hundreds of millions of dollars in revenue.

But these games that sell in the millions are dwarfed in terms of ubiquity by Solitaire and Minesweeper on Windows-based PCs, and much more recently by games played on smartphones or over the internet. Angry Birds has now been downloaded an incredible 300 million times, earning a fortune for its developer, the Finnish company Rovio Mobile; Farmville and other Facebook games created by developer Zynga are played by 270 million people; and more than 10 million people pay $10 every month to play World of Warcraft, the massive multiplayer online role playing game published by Activision Blizzard, Inc.

Clearly, video gaming has developed into a very substantial business. In 2011, revenue for video and computer gaming will be more than $74bn. Some estimate that this could rise to $112bn by 2015, with most of that growth coming from casual games.

Meanwhile it is fitting that Nolan Bushnell is still involved. Having founded 20 companies since Atari, at the time of writing he is running Anti-Aging Games, which produces games scientifically developed to stimulate the brain.

24

Just-in-Time inventory management

When: 1973

Where: Japan

Why: Just-in-Time transformed the way all significant global manufacturers operate

How: An American supermarket's unusual procurement strategy sowed the seed that grew into lean manufacturing

Who: Taiichi Ohno

Fact: When Toyota implemented Just-in-Time at its Nagoya plant, its response times fell to a single day

J ust-in-time (JIT) inventory management is the philosophy at the core of the Toyota Production System (TPS) developed between 1948 and 1975 by Shigeo Shingo, Eiji Toyoda – and Taiichi Ohno. Toyota's methods revolutionised manufacturing in Japan's automotive industry and, with cultural and sector variations, were eventually adopted throughout the world, first in discrete manufacturing (the sort that produces physical objects), then in process manufacturing and finally in business generally.

The background

Until the arrival of JIT, inventory management depended heavily on forecasting, often weeks ahead. The number of units that the business expected to produce would be worked out based on orders in hand, market forecasts and last year's performance. Really lazy companies would just follow the last of these, and act retrospectively if they sold more or less than anticipated.

To support this model, manufacturers had to order enough parts ahead of time to meet the forecast. It was not uncommon for the stores to contain three months' inventory ready to meet production targets. If the stock ran out before the next scheduled order date, more would have to be sourced in a hurry; if there was too much, it would remain on the shelves.

In 1950 Toyota's managing director, Eiji Toyoda, and executive Taiichi Ohno led a delegation to the USA, visiting a Ford plant in Michigan. They were impressed by its scale, though not by its efficiency. The production process resulted in large quantities of surplus inventory, and an uneven flow of work. Furthermore, because the quality checks were left until the end of the process, many cars were being sent back for re-working.

Henry Ford had envisaged lean principles in 1923 when he wrote: 'We have found in buying materials that it is not worthwhile to buy for other than immediate needs. If transportation were perfect and an even flow of materials could be assured, it would not be necessary to carry any stock whatsoever ... That would save a great deal of money, for it would give a very rapid turnover

Taiichi Ohno, the founder of JIT.

and thus decrease the amount of money tied up in materials.' But Ohno found no evidence that this principle was being applied in Ford's factories.

On the same trip he visited the self-service grocery chain Piggly Wiggly, where he found a fully functional JIT programme in practice for the first time. Replacement stock was ordered only when existing quantities reached a critical level. He went back to Japan convinced of the need for a production system that was pulled by real orders, rather than pushed by sales forecasts.

The JIT process, defined by Ohno and adopted at all Toyota plants, hinged on: 'Making only what is needed, when it is needed, and in the amount needed.' Quality was made an integral part of the process by eliminating waste, inconsistencies and unreasonable requirements on the production line.

An approach that had initially been regarded with suspicion in the West, due to lazy, xenophobic stereotypes about Japan ... gradually became accepted because it improved productivity and profitability, and identified waste.

The JIT system, as implemented by Toyota, dictated that a production instruction must be issued to the beginning of the vehicle production line as soon as a vehicle order was received. The process required a close working relationship between the assembly line and the parts-production process. Any parts used on the assembly line had to be replaced from the parts-production stores, in exactly the same quantity; the parts team were instructed to replenish the assembly stock, and no more – surplus production was eliminated.

The system wouldn't work without the collaboration of parts suppliers and sub-assembly manufacturers – so Toyota began using 'Kanban cards' – product request forms that would be passed back through the supply chain as soon as stocks ran out. Although the Kanban required the suppliers to change their modus operandi, they clearly stood to gain from it – as the whole object of JIT was to produce more vehicles, bringing improved returns all the way down the supply chain.

Following the global oil crisis of 1973, which triggered soaring fuel prices and falling supplies, Toyota recovered markedly faster than its peers; its resilience gained international respect, and manufacturing businesses began to consider JIT as a viable commercial approach. An approach that had initially been regarded with suspicion in the West, due to lazy, xenophobic stereotypes about Japan's love of order and control, gradually became accepted because it improved productivity and profitability, and identified waste.

More and more companies began to revamp their inventory management strategies, with JIT as the guiding influence. In 1984, General Motors formed a joint venture agreement with Toyota for the production of small cars in the USA – buying into JIT in the process. In 1986 Richard J. Schonberger, a respected authority on JIT, wrote that more than 100 leading American firms had tried out the system – including famous names such as Intel, Motorola and Campbell's Soup.

Commercial impact

Today, thousands of companies around on the world rely on the ideas initially conceived by Taiichi Ohno. In 2005, a group of senior manufacturing executives were asked about their approach to inventory management; 71% said they used JIT. Over recent years, the principles of JIT have evolved into a new, Western concept known as 'lean manufacturing', which shares the same commitment to paring costs down to their absolute minimum by imposing rigorous control over the production process, and cutting back on inventory at every stage. Lean manufacturing is now used across a multitude of industries; many companies believe it has played a key role in helping them weather the recession.

One of JIT's most ardent proponents has been IT hardware giant Dell, which has stripped back its production operation to almost completely eradicate inventory. Although Dell puts nearly 80,000 computers together every day, it doesn't own or rent a single warehouse, and carries no more than two hours' worth of inventory at any one time. Experts believe this approach has brought genuine advantages; in fact, it is thought that Dell's cost-effective production process allows the company to undercut its rivals on price by up to 15%. Dell claims that, because its parts do not arrive until the eve of assembly, they are around 60 days newer than the components of IBM and Compaq machines; this can bring a profit advantage of up to 6%.

Lean manufacturing is now used across a multitude of industries; many companies believe it has played a key role in helping them weather the recession.

Meanwhile, Toyota has been able to establish a position of global strength in the automotive market, based on its ultra-efficient manufacturing process. In 1998 it was found that Ford and GM were taking 50% more time to manufacture a car than Toyota. While GM was not even making a profit on several models, Toyota was making profits stretching into billions of pounds. In an attempt to close the gap, GM began a concerted campaign to emulate Toyota's production

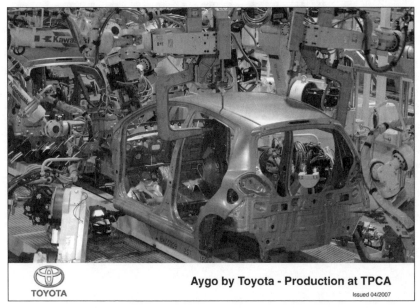

Aygo by Toyota - Production at TPCA

Issued 04/2007

TOYOTA

JIT inventory management is still used today.

methods, based on JIT. By 2007, GM had implemented the new system in 90% of its plants around the world; the company claimed that this brought cost savings of around $9bn.

What happened next?

Toyota promoted Taiichi Ohno to its board in 1954. He became managing director in 1964, senior managing director in 1970, and executive vice president in 1975. He retired from Toyota in 1978 and died on 28 May 1990 in Toyota City, Japan. Ohno's influence was felt well outside the field of manufacturing, where JIT has played a huge role in improving back-office functions, sales and customer relationship management.

25

Magnetic Resonance Imaging (MRI)

When: 1970s

Where: USA and UK

Why: It has allowed doctors to see detailed images of the inside of patients' bodies without surgery

How: Three inventors separately pursued a way to improve diagnostic medicine

Who: Paul Lanterbur, Sir Peter Mansfield and Raymond Damadian

Fact: The MRI market is worth in excess of $5bn

D octors have always been fascinated with what is taking place inside patients' bodies. While diagnoses can sometimes be deduced by examining a patient externally, at other times only the internal workings of the body hold the key to what is wrong. Surgery is expensive, intrusive and potentially dangerous, so being able to see inside the body without physically cutting a hole had long been an ambition for the medical profession.

The earliest breakthrough came in the late 19th century with the discovery of X-rays and the subsequent machines. However, these could only really be used to look at bones, and something more subtle would be required to spot other illnesses. Magnetic Resonance Imaging (MRI) became that critical tool. The MRI uses a strong magnetic field to produce a reaction in the nuclei of cells. These reactions are scanned, creating a detailed internal picture of the body and enabling doctors to see on a screen such things as tumours and cancers and make a definitive diagnosis. Over the past 30 years, the MRI machine has become a genuine life-saver.

The background

During the 1950s, a British scientist called Peter Mansfield began to work in the field of nuclear magnetic resonance (NMR). Others before him had seen the potential of the area as a way to study the molecular structure of chemicals. Mansfield was tasked with creating a machine that would harness this nascent technology to create a practical scientific application. He designed a portable spectrometer powered by transistors that could be used to scan chemicals and provide key information about them. Mansfield thought that if such a device could be safely used on a living human being, then it would have remarkable implications for medical science. Shortly after his work on NMR, Mansfield embarked on a path of study that enabled him to make key discoveries leading to the creation of the MRI machine.

Meanwhile, on the other side of the Atlantic, Paul Lanterbur was heading in a similar direction. He was drafted into the army in the 1950s, but was allowed to pursue his scientific pursuits working on an early NMR machine. He published several papers on NMR and, like Mansfield, pursued the idea of a machine that could be used to work on a human body.

By the 1970s, the two men were making breakthroughs independently. Lanterbur was working at Stony Brook, part of the State University of New York. Here, he hit upon a number of ideas that, by the 1970s, helped to create the first MRI scanner. The machine focused high-powered sound waves on specific areas of the body, which would agitate the cells. The excited cells would send out radio waves that could be scanned and used to produce a detailed internal picture of the body. Lanterbur introduced the idea of using gradients in the magnetic field to pin-point the origin of these radio waves. Mansfield, meanwhile, working at the University of Nottingham, discovered how the data made from a scan could be analysed. Both men built early prototypes of MRI machines in the early 1970s. Mansfield actually managed to create a machine that was able to successfully scan a human hand.

1977 – Drs Raymond V. Damadian, Lawrence Minkoff and Michael Goldsmith (left to right), and the completed Indomitable, the world's first MRI scanner.

Professor Sir Peter Mansfield – his pioneering work on MRI was recognised with the Nobel Prize for Medicine in 2003.

> Both Mansfield and Lanterbur should have become very wealthy as a result of their efforts ... However, only Mansfield was able to successfully patent his work. Lanterbur tried to, but was not supported by Stony Brook […] this later proved to be a terribly unwise decision.

As the research emerged, commercial interests in the creation of a fully functioning MRI machine became highly active. The aim was to create a machine that could scan an entire human body and be deployed in a hospital. Both Mansfield and Lanterbur should have become very wealthy as a result of their efforts, as the medical industry was likely to spend millions on this ground-breaking research. However, only Mansfield was able to successfully patent his work. Lanterbur tried to, but was not supported by Stony Brook, which believed that the cost of taking out a patent would never be recouped – this later proved to be a terribly unwise decision.

By the late 1970s, however, another man was claiming credit for the MRI scanner. Dr Raymond Damadian had written a number of papers on MRI and had successfully lodged patents. In 1977, he also created the very first MRI scanner that could conduct an entire body scan. Damadian dubbed the 1½ ton machine 'The Indomitable' and a year later set up the Fonar Corporation to sell MRI scanners based on his successful prototype.

Commercial impact

As the 1980s began, big business moved into the production of MRI scanners for commercial sales. However, controversy was rarely far away because Damadian pursued legal action against many companies for breaching his patents. Most famously, General Electric had to pay out $129m.

The legal fracas probably slowed the spread of the MRI machine into hospitals, but eventually the matter was settled. Lanterbur and Damadian engaged one another in a battle of words and insults over who was to take the credit for MRI, a matter that they never resolved.

Once the legal barriers were overcome, the medical establishment rapidly invested in the machines. This meant the money made could go back into research and development and be used to further improve the products. By the mid-1980s, the speed at which the machines could produce images had radically improved, and today the scans work in real time. MRIs can also produce highly detailed images of the brain and have helped to advance medical science considerably. By 2002, it was estimated that 22,000 MRI cameras were in use and over 60 million scans had been conducted. Today the MRI scanner market is worth $5.6bn globally and some of the world's biggest names, such as General Electric, Siemens and Philips Medical Systems, are all involved in their production and deployment.

By 2002, it was estimated that 22,000 MRI cameras were in use and over 60 million scans had been conducted.

What happened next?

The controversy over MRI rages on to this day and was notably inflamed in 2003, when the Nobel Prize in Physiology and Medicine was awarded to Paul Lauterbur and Sir Peter Mansfield for their contributions to MRI. Nobel Prizes can be awarded to up to three people. However, the judges decided not to include Damadian. He was furious at the decision and took out full page advertisements in the New York Times, the Washington Post and the Los Angeles Times to make his displeasure known. He and Lanterbur continued the war of words that had begun years earlier, but it was only the latter who took the distinguished prize.

The story of MRI shows just how controversial research and development can be, especially when scientists are all heading in the same direction. It also demonstrates the importance of patenting your research. Damadian never got a Nobel Prize but did become very rich. Lanterbur got the credit but never made the fortune he might have done. Mansfield, by successfully realising his vision and attaining a patent, enjoyed both the wealth and the acclaim.

26
VHS

When: 1973

Where: Japan

Why: The VHS transformed the cinematic experience, allowing people to watch movies and programmes at a time and place of their convenience

How: JVC and Sony embarked on a battle for control of the emerging video market, which would ultimately bring victory for VHS by the mid-1990s

Who: JVC

Fact: *The Lion King* is the biggest-selling VHS release of all time

Today's world, brought to us in high definition (with surround sound, fast navigability and bonus features as standard), can make the era of the Video Home System (VHS) seem to be light years ago. But in the 1970s, the old black video cassette, known to many simply as VHS, was a significant step forward in home entertainment, for the first time ever giving consumers enormous control and choice over what to watch at home.

The home video market created a lucrative new revenue stream for the major film studios, and allowed film buffs to indulge their hobby by watching hundreds of different movies in the comfort of their own living rooms. Some now believe that the rise of VHS shaped the work of directors such as Quentin Tarantino, whose film obsession is demonstrated by the hundreds of off-beat and obscure references he weaves into his releases. Furthermore, it is arguable that 1980s classics such as *Ghostbusters*, *Indiana Jones* and *Back to the Future* would never have earned the iconic status they enjoy today without the millions of video cassettes that took them into homes around the world.

The background

It is generally thought that VHS technology dates back to the mid-1970s, but in fact it relied on a series of advancements that began over two decades earlier. Work on the video cassette recorder (VCR), which played VHS tapes, began in the post-war period as electronics firms woke up to the potential for a machine that could replay films and television programmes over and over again, taking the movie experience into the viewer's living room.

Attempts to create a commercially viable video recording device were led by a young firm called Ampex, whose work on a video tape recorder began as early as 1951, long before most consumers were able to record and play back music. Their product, based on four rotating heads, was first unveiled a year later, beginning a period of piecemeal improvement. Although money was tight and progress was hampered by frequent delays and suspensions, the development team was able to make a series of measured enhancements to picture and sound, and by the spring of 1956 the major broadcasting networks were queuing up

Ampex's founder Alexander M. Poniatoff with the early VR1000

to have a look at Ampex's video-tape recording machine. By November, CBS was broadcasting programmes from video tape.

The Ampex innovation sparked a wave of imitations – indeed the BBC experimented with a similar system, called VERA, during the late 1950s. But these early breakthroughs, though undoubtedly significant, were never going to translate into commercial success; the machines were massive (the early Ampex machine, known as the Quad, weighed almost half a ton), the tape heads ran out quickly, and there was no way to freeze images.

The VHS was fairly nondescript in appearance. But it was cheap and simple, both crucial advantages in the battle that was to follow.

Slowly, the equipment began to get smaller – a cluster of Japanese firms, including Sony, JVC and Matsushita, took a long look at the Ampex machine, and miniaturised its components using revolutionary recording-head mechanisms and solid-state electronic circuits. By 1963, Sony was able to introduce the PV-100, the first compact video recorder.

Although the early models were blighted by high prices, low picture quality and reel-to-reel formats that lacked convenience for the end user, the major manufacturers continued to push forward, racing each other in an attempt to master video-recording technology. Sony launched the first video cassette, the U-matic, in 1969, and began commercial production of the necessary recording technology in 1971. The machine, which relied on ¾-inch magnetic tape, failed to catch on beyond a hard core of companies and education providers; its high cost (over $7,000 in 2007 dollars) was beyond the purse of all but the wealthiest families. However, it served as a crucial template for the video-recording equipment that was to follow.

At around the same time as work was beginning on the U-matic, Sony began a period of collaboration with Matsushita Electric and JVC, the company that had produced the first Japanese television, in an attempt to agree a common video standard. But within months the collaboration effort had broken up, creating a feud which would simmer for years to come. Sony began work on Betamax, while JVC started exploring the development of an alternative originally known as Vertical Helican Scan, or VHS. However, the latter project was side-lined shortly afterwards as falling revenues forced JVC to reassess its priorities. Nevertheless, the project's two senior engineers – Yuma Shiraishi and Shizuo Takano – gamely kept going, working in secret.

By 1973, Shiraishi and Takano had created a prototype, and after three years of further research, testing and refinement, JVC unveiled the first VCR to use

VHS, the Victor HR-3300. Sales began in Japan the following month, and the USA gained its first VCR machine, the RCA VBT200, the following August, with units costing around $1,000.

Much of the early growth of VHS was triggered by Matsushita, which produced more than half of all Japanese VCRs by 1978. The relatively simple design of the VHS cassette changed little before it became obsolete almost three decades later. The mechanics were based on spools of magnetic tape encased within a plastic clamshell. The tape was pulled by a conveyer belt of guides and rollers, and a helical scan video head rotated against it, creating a two-dimensional image. Compared to the silvery, shimmering style of the DVD that ultimately replaced it, the VHS was fairly nondescript in appearance. But it was cheap and simple, both crucial advantages in the battle that was to follow.

Commercial impact

The arrival of VHS heralded an instant boom in sales – the value of Japan's VCR export industry doubled between 1976 and 1977, and had doubled again by 1983. But before it could gain control of this burgeoning sector, JVC had to win two key battles: with the film studios, which wanted to ban home ownership of VCRs as a violation of copyright; and with Sony, which had brought Betamax to market in 1975. Ultimately, through a process of gradual assimilation, expansion and self-perpetuating growth, JVC overcame both opponents.

The VHS-Betamax battle pitted speed against quality. While the image produced on Betamax was slightly better, a VHS cassette could rewind and fast-forward much more quickly, thanks to a simpler unthreading system. It could also provide far more tape time – up to nine hours – with a maximum of three hours' recording time, compared to just an hour on Betamax – which, crucially, wasn't enough to capture the full length of an average movie.

The VHS Group agreed on a common European standard, which yielded export agreements with a network of European distributors.

JVC was quick to find allies in its battle with Sony, reaching out to its competitors in an attempt to spread its standard. By the end of 1976, Sharp and Mitsubishi had adopted the VHS model, while Hitachi had begun marketing JVC's video cassette machines. The following year the VHS Group agreed on a common European standard, which yielded export agreements with a network of European distributors.

More and more manufacturers and distributors entered into partnership with JVC, widening the production operation and bringing economies of scale, which were enhanced by the simple, cost-effective design of the VHS. Key components such as polystyrene and polypropylene could be procured cheaply, and would last for up to 20 years. Consumers knew they were getting a durable, reliable product which could stand everything from long-haul transit to a child's rough handling.

Jack Schofield, writing about the VHS–Betamax battle in the *Guardian* back in 2003, outlined the advantages of VHS in simple terms, saying: 'VHS offered a bigger choice of hardware at lower cost, the tapes were cheaper and more easily available, [and] there were a lot more movies to rent.' While this was true, Sony made bigger technological strides than JVC during its scramble for control of the video market. Indeed, Betamax was first with a number of features that have become modern-day viewing staples, such as hi-fi sound, wireless remote control and digital freeze-framing. But the reliable, inexpensive design of the VHS offered greater appeal to manufacturers, and the design was open to everyone.

JVC maintained a loose licensing framework, which allowed numerous studios and cassette manufacturers to go into VHS production. In contrast, Sony's founder, Akio Morita, has since conceded that licensing problems with other companies slowed the growth of Betamax considerably. Thus, by the end of the 1970s, VHS was clearly winning the battle. In 1978 it overtook Betamax in overall global sales and by 1980 it commanded 70% of the US market. By 1984, 40 manufacturing companies were using VHS, while only 12 had opted to go with Betamax. VHS's open approach to partners and distributors had brought market dominance all over the world.

Indeed, in 1983, VHS outsold Betamax by at least 40% in the USA, Japan, West Germany, France, Italy and the UK. For Sony, which had fired the first shots in the video cassette conflict back in the early 1970s, there was now nowhere to hide. In 1988, the Betamax manufacturers finally raised the white flag and began producing VHS video recorders themselves. (And Sony had learned the significant lesson: that software is crucial to success in selling entertainment hardware. It would later buy its own movie studio, and eventually computer games developers, in order to avoid repeating the Betamax mistakes.)

The battle with the studios was far less taxing. The movie producers initially opposed the advent of home video entirely, concerned that it would reduce cinema audiences. But soon the likes of Universal Studios decided to harness the VHS, and began producing video copies of the movies they made. This gave rise to a new and substantial revenue stream. According to the *Movies and Entertainment: Global Industry Guide*, released in April 2011, the home video sector now accounts for more than 50% of the global movies and entertainment market.

This growth was driven by the plethora of video rental stores that opened throughout every country in the western world – creating a vibrant small new retail sector. Blockbuster was the most successful company to take advantage of this, though it closed offline stores and has launched an online video streaming service. By creating such a significant new revenue source for film producers, the VHS enabled thousands of movies to find commercial viability – movies that were too niche or small for cinemas (which could show only a relatively small number of new films in any year) could suddenly reach an enormous customer base; the 'straight to video' segment had been created.

After years of growth, VHS sales ultimately peaked in the mid-1990s – when *The Lion King* sold more than 55 million copies worldwide. But, even as sales were hitting their high point, the era of the VHS was already coming to a close.

What happened next?

JVC tried to move with the times – a new format, SuperVHS, was rolled out, offering superior picture quality. While Sony may have been first to deliver a series of ground-breaking modifications, VHS caught up with each of them in time. But the emergence of the digital versatile disc, or DVD, brought VHS's period of pre-eminence juddering to a halt.

As the pace of technological change gets ever quicker, it's safe to assume that VHS will never make a commercial comeback – but its place in the hearts of millions ... will never be erased.

The first commercial optical disc storage device had been the Laserdisc, which had launched in 1978 but was poorly received in Europe, North America and Australia. However, the DVD, its descendant, was an altogether different proposition; the analogue footage on VHS could never be as sharp as the digital material provided by a DVD, and the new dual-layer, ultra-lightweight discs were capable of storing around eight hours of high-quality video – or 30 hours of VHS-quality footage.

When DVDs first came to market in the late 1990s, their impact was limited by the cost of the machinery required to play them. But, by the dawn of the new millennium, prices had dropped significantly; a US consumer could pick up a DVD player for less than $200. Furthermore, by the end of the 1990s most of the major studios had begun to support DVD technology, and around 100 movies were being newly released via this channel every week.

It was in June 2003 that DVD sales began to outstrip VHS in the USA. The last mass-produced VHS release was *The History of Violence*, which went to video in 2005. Thereafter the commercial video production of movies stopped. Around 94.5 million VHS machines were still in use across the USA at this time, according to the Washington Post, but in October 2008 the last major shipment of video cassettes in the USA brought the VHS era to a definitive close.

In recent years the home video market has stalled. A report from Strategy Analytics, released in May 2010, predicted that global home video sales will fall to $48.1bn by 2013 – almost $7bn down on their 2009 return. Meanwhile, the VHS has fallen into total obsolescence. Yet, in recent times a trend has appeared that seems to be rooted in a nostalgic revival. VHS enthusiasts have preserved the product's image in all manner of weird and wonderful ways – several websites are now even offering hard-bound notebooks that resemble a VHS tape!

As the pace of technological change gets ever quicker, it's safe to assume that VHS will never make a commercial comeback – but its place in the hearts of millions of 1980s children will never be erased.

27
The barcode

When: 1973

Where: USA

Why: The automation of the retail process saw a transformation of shopping

How: Inventors wanted to devise a system whereby product information could be automatically read during check-out

Who: Joseph Woodland and Bernard Silver

Fact: The very first item scanned by the standardised system in 1974 was a pack of Wrigley's chewing gum

Today, barcodes are so commonplace that it's almost impossible to imagine the retail environment without them. We don't give a second thought to shop assistants scanning a can of soup, a new pair of shoes or a bunch of bananas; we even do it ourselves on self-service check-outs, or buy iPhone apps with barcode scanning capabilities. But to get to this point, the concept has gone through many different guises and incarnations.

It took three decades to turn a theory into a fully functional reality – and much, much longer for it to become the retail cornerstone that it is today.

The background

Before the barcode, shopping for items required the cashier to manually enter in the price of each item, a slow and tedious process with enormous capacity for error. It also meant that record-keeping and writing of receipts had to be performed manually, so keeping track of stock was a painstaking task. The industry was crying out for change.

Back in 1932, a bunch of Harvard students offered early hope by creating a system of punch cards. However, it wasn't until the late 1940s that Bernard Silver, a graduate student at Drexel Institute of Technology in Philadelphia, overheard the boss of a local food chain asking one of the deans if it would be possible to research a system that could automatically read product information during the check-out process. Excited by the prospect of creating a game-changing technology, Silver repeated the conversation to his friend Norman Joseph Woodland, a 27-year-old graduate and teacher at Drexel. They began work immediately, and the long process that led to the creation of the barcode was under way.

The duo certainly came up with a few red herrings. Their first working prototype, based on light-sensitive ultraviolet ink, was soon found to be both expensive and ineffective, because the ink faded too quickly. So they went back to the drawing board; Woodland quit his post at Drexel and moved into his grandfather's apartment in Florida, where he continued beavering away. The next

The use of barcodes transformed the grocery industry.

version was inspired by Morse code: 'I just extended the dots and dashes downwards and made narrow lines and wide lines out of them', he reportedly said.

But creating the machines to read the data wasn't so simple. Eventually, Woodland decided to adapt the movie sound system technology dreamt up by Lee de Forest in the 1920s. Convinced he'd hit the jackpot this time, Woodland headed back to Drexel. He decided to replace his wide and narrow lines with concentric circles, so that they could be scanned from any direction; this became known as the bull's-eye code.

Shortly after, on 20 October 1949, the two filed for the 'Classifying Apparatus and Method' patent. The inventors described their creation as 'article classification ... through the medium of identifying patterns' – referring to the iconic barcode pattern of vertical lines that we all know so well these days.

While the patent was pending (it wasn't granted until 7 October 1952), Woodland took a job at IBM, where he unsuccessfully attempted to persuade his bosses that IBM should get involved in the development of his invention. Finally running out of steam, the two inventors sold their patent to Philco, a pioneer in early battery, radio and television production in 1952; Philco then sold it on to RCA (the Radio Corporation of America) in the same year.

Meanwhile, David Collins was developing an invention of his own. As a student, he had worked at the Pennsylvania Railroad, where he noticed the complexity of keeping track of train cars. After some investigation, it became apparent that some sort of coded label would be the easiest and cheapest way of solving the problem.

The labels Collins eventually came up with were not what we now think of as barcodes: rather than being formed of black lines or rings, they used groups of orange and blue stripes made of reflective material. But Collins did have the foresight to figure out that there were many applications for automatic coding beyond the railroads.

He pitched the idea to his current employer, Sylvania, but short-sighted bosses rejected his proposal because of a lack of funding. So Collins promptly quit and set up his own company, Computer Indentics. In 1969, Computer Indentics quietly installed its first two barcode systems – one at a car plant and another at a distribution facility.

But it was the grocery industry that had most to gain from the adoption of barcodes. At an industry event in 1971, RCA, which had been pouring ever-increasing resources into the project, demonstrated a functional bull's-eye code system. Having got wind of the idea's considerable popularity (and worried that it could be missing out on a huge market), IBM somewhat belatedly decided that it wanted a piece of the action.

The adoption of the Universal Product Code, on 3 April 1973, transformed barcodes from a technological curiosity into an idea that would change the retail business for good.

Handily, Woodland, the barcode's original inventor, was still on IBM's staff. His patent had expired, but he was transferred to IBM's facilities in North Carolina, where he played a prominent role in developing the most important version of the technology: the Universal Product Code (UPC). It soon became apparent that an industry standard was needed (multiple versions of the barcode would be worse than useless), and while RCA continued to push its bull's-eye version, the technically elegant IBM-born UPC was the one ultimately chosen by the industry. The adoption of the UPC, on 3 April 1973, transformed barcodes from a technological curiosity into an idea that would change the retail business for good.

By June 1974, all the groundwork was in place – the testing carried out, proposals completed and standards set – and a pack of Wrigley's chewing gum became the first retail product sold with the help of a barcode scanner. Today, the pack of gum is on display at the Smithsonian Institution's National Museum of American History.

Commercial impact

After decades of development and billions of dollars in investment, barcodes and the accompanying scanners had now become a practical reality. That's not to say that they were wholeheartedly embraced by the retail community; on the contrary, the use of scanners grew slowly at first. The system would only start paying off when around 85% of all products carried the codes. But when this target was met, in the late 1970s, take-up rates rocketed.

It's little wonder that it caught on, because the benefits of barcodes are plentiful. The first advantage for retailers is speed: a barcode can be scanned in less time than it takes for a keyboard operator to make a single keystroke, and mistakes are much less common. It's cost-effective, too: research conducted by consultancy McKinsey back in 1970 predicted that the industry would save $150m a year by adopting the systems. And it also gives retailers much more accurate data about their customers – without exposing themselves to lost business and higher operating risk. So there's no doubt that barcode technology was a big step forward for the retail industry.

A barcode can be scanned in less time than it takes for a keyboard operator to make a single keystroke.

An ABI Research report in early 2011 estimated that the market for barcode scanners, which includes both fixed and handheld devices, was worth $1.4bn in 2010. It anticipated that 2011 would see growth stall somewhat, amid financial turmoil and a slump in the retail sector, with revenues increasing by just $0.1bn during the year for leading providers of devices such as Motorola, Psion Teklogix, Nordic ID, AWID, Intermec, and Convergence Systems Limited.

It is expected that manufacturers will drive additional revenue from the increasing prominence of 2D versions of the traditional barcode – ones that consist of not just straight lines, like the original version, but of squares, dots, and other geometric patterns.

What happened next?

Barcodes such as the UPC have become a ubiquitous component of modern-day life. And not just in the retail business; the technology has been slowly creeping into other areas of life in a way that Woodland and Silver could never have imagined. Scientists have even put miniature barcodes on bees to track their mating habits, while the codes also appear on concert tickets, hospital ID bracelets, business documents and even marathon runners' bibs.

Another evolution of the classic barcode is the RFID (Radio Frequency Identifier) tag, which is also increasingly common; they're stuck to the covers of books and CDs to prevent them from being stolen, and are used extensively in warehouses and other areas of the supply chain to keep track of stock. Some countries even employ RFID to identify lost pets.

Scientists have even put miniature barcodes on bees to track their mating habits.

The appearance of barcodes has evolved, too. The UPC version remains the most dominant and recognisable – but like many technologies, it's in danger of being overtaken by newer, more advanced versions. The new 2D versions, for example, are being used on airline boarding passes throughout the world, supplanting the traditional magnetic design. Unlike magnetic strips, 2D barcodes can be printed on any kind of paper, or can even be scanned directly from a passenger's mobile phone. This not only saves time for passengers, but also money for airlines because they need fewer check-in staff – an all-round win-win.

One particular form of the 2D code that has captured the attention of the public, technology companies and marketing agencies alike is QR (Quick Response). Created by Toyota in 1994, QR was originally used to track vehicle parts. But it has since found more glamorous uses, having been seen on the cover of a Pet Shop Boys' single and in PlayStation 3 games.

And what became of the founders? Sadly, Silver died in 1963 before the value of his invention had been truly realised. As for Woodland, having sold the patent, it's unlikely that he ever became rich from his invention – although he can take some solace in the fact that he was awarded the 1992 National Medal of Technology by President Bush. He is thought to be now retired, safe in the knowledge that, as a result of his work, neither retail nor logistics will ever be the same again.

28

Electronic point-of-sale (EPOS) technology

When: 1973

Where: USA

Why: Transformed the stock control and sales analysis process for retailers, enabling substantial growth

How: IBM's EPOS system precipitated a revolution in customer service

Who: IBM research

Fact: IBM's gross income passed $10bn in the year it launched its first EPOS machines

Standing in a busy supermarket, fashion boutique or fast-food restaurant today it is hard to imagine how businesses coped before the introduction of electronic point-of-sale (EPOS) technology. Certainly the speed and ease of customer service would be impossible.

The forerunner of all the EPOS technology used in today's stores and service industries was the IBM Store System, launched in the early 1970s. IBM's influence on EPOS continues to this day but the industry is far more competitive than when the Store System was launched.

Barriers to entry have been removed as EPOS innovation has shifted from hardware to the software that runs on it. Hosted on remote servers with real-time data back-up, hundreds of EPOS software options are available and affordable for even small business owners. Back in the 1970s, the breakthrough IBM Store Systems were priced at the top end of the market, affordable only for larger, more profitable businesses.

The background

Before the launch of the first EPOS systems, retailing involved lots of manual tasks – such as checking stock, and entering each item a customer wanted to buy into the till by hand. Not only was this time-consuming and complex to manage, but human error also substantially increased the likelihood of mistakes.

The lack of real-time correlation between sales and stock meant that retailers would often run out of certain items if they proved popular. Business trends over any concerted time period were hard to identify, let alone evaluate, and missing stock was almost impossible to account for, making theft by employees or customers more difficult to trace.

In the same year that IBM brought out its Store Systems, the first barcode scanner was in development.

The IBM 3650 and 3660 Store Systems, launched in 1973, precipitated a new era in customer service and retail. At the time, IBM was the pre-eminent force in researching new technology for automating business processes – several Nobel Prize winners were associated with the company's research work – and the IBM Store Systems grew out of this research. They were predicated on a central mainframe computer, which communicated with a series of 'dumb terminals' – computers that lacked any processing capacity and were used as cash registers. This technology was undoubtedly primitive compared to the functionality offered by EPOS systems today; however, it still represented

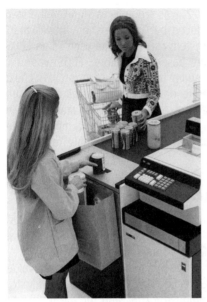

IBM 3663 Supermarket Terminal.

a huge leap forward for the owners of retail businesses. In the same year that IBM brought out its Store Systems, the first barcode scanner was in development (see the chapter on barcodes for more information), and a set-up resembling the type recognisable in retail and service industries today started to take shape.

Rivals firms such Regitel, TRW, Datachecker and NCR competed with IBM for market share throughout the 1970s, but in reality, only these few large and well-established companies could afford to push forward the capabilities of EPOS. Other technological developments did alter the possibilities and expectations of EPOS, however. McDonald's use of an early type of microchip in its tills, for example, allowed staff to serve more than one customer at a time. And developments in stand-alone credit-card devices meant that credit-card processing facilities were gradually incorporated into the EPOS technologies.

In 1986 a deli owner called Gene Mosher made a crucial breakthrough by creating the first-ever touchscreen EPOS device; this triggered a wave of innovation, as hundreds of small companies began creating their own versions of EPOS, tailored to specific types of retailing. IBM rolled out an updated version of its Store System, in an attempt to keep up with the quick-witted independents, and then Microsoft got involved by creating IT Retail – the first point-of-sale software compatible with Windows.

The new system, which launched in 1992, was arguably as significant a step forward as the touchscreen device produced by Gene Mosher. Thanks to IT Retail, a PC with a Microsoft operating system could now be used as an EPOS device; having started out as a hardware-based technology, EPOS was now being rolled out in software form. This meant that many more retailers could access the technology – it's far simpler to download and install a software package than to purchase and put together a bulky piece of machinery. In the late 1990s an industry-wide standardisation campaign completed the transformation of EPOS from an expensive, largely inaccessible rarity to a staple business tool.

Commercial impact

The take-up of EPOS has been spearheaded by supermarkets, which required technology capable of managing multiple outlets, directing stock from several warehouses and providing a quick, painless customer experience.

By the late 1990s, around 89% of the UK grocery industry's retail sales value was passing through outlets with EPOS technology. For the music industry this figure was 80%, for the bookseller industry 70%, and for the magazine industry 46%.

The implementation of EPOS has brought multiple benefits. A survey commissioned by Britain's Institute of Grocery Distribution in the late 1990s, asking financial directors from across the industry about the impact of EPOS technology, found that 100% of companies had seen cost-reduction benefits. Almost 75% had witnessed greater labour efficiency, 63% had benefited from improved product availability and 55% had improved their security through point-of-sale systems.

The take-up of EPOS has been spearheaded by supermarkets ... managing multiple outlets, directing stock from several warehouses and providing a quick, painless customer experience.

However, the extent of the benefit provided by EPOS varies considerably from industry to industry. As noted in a report released by Retail Banking Research in August 2011, large chains have adopted EPOS far more enthusiastically than have independent outlets. Thus the retail sector, which is dominated by chains, has gained considerably from EPOS; in the hospitality industry, which remains extremely fragmented, take-up has been far lower, with many businesses still relying on more basic equipment, such as cash registers.

EPOS manufacturers have tended to struggle in recent years – in fact global sales fell from 1.62 million terminals in 2007 to 1.24 million in 2010, before recovering to reach 1.38 million the following year. IBM, having been tested by Microsoft and a cluster of nimble independents in the 1980s and 1990s, appears to have re-gathered its strength and is currently the global EPOS market leader, accounting for around 22% of global sales. Asia is the largest market, its growth driven largely by increased demand from the burgeoning Chinese economy.

What happened next?

EPOS systems today fall into two categories: Modular and Integrated. As the name of the latter suggests, the display unit and computer are integrated, whereas in a Modular System a computer is connected to peripheral devices such as scanners, printers and the display device. The real diversification in the EPOS marketplace today is not in hardware, however, but in the software used on the systems. There are hundreds of software vendors operating today, a far cry from a behemoth-like IBM exerting hegemony over the market as it did in the 1970s. Many large businesses now run their EPOS via a client–server model in which there is a dedicated database server to which all other stations talk for retrieving and storing data.

The dramatic growth of internet use and bandwidth over the past decade has made this type of remote hosting commercially viable, even for SMEs. The internet, and particularly online retail, has also changed expectations of what is to be delivered by EPOS software. Software is now expected to integrate a company website, monitoring what has been sold online and adjusting stock orders accordingly, reducing stock holding and freeing up cash flow, something of paramount importance to businesses today.

EPOS software will track not only sales but also the performance of staff. Perhaps most importantly, software can be tailored to suit the specific needs of a company. This is the main change, compared to the IBM Store Systems of the 1970s. Companies investing in EPOS software today are investing in a business relationship rather than specific hardware. Machines are still needed to run the software, and mainframes are increasingly important in data centres for ensuring real-time back-up, but it is the developers and administrators of the software that are the backbone of the EPOS industry today.

29

The Global Positioning System (GPS)

When: 1973

Where: USA

Why: The technology transformed navigation and created a new market

How: The US government permitted civilian use of military-grade satellites

Who: Roger L. Easton

Fact: The GPS market is thought to be worth $75bn worldwide

The technology that enabled the development of the Global Positioning System (GPS) is now a part of our everyday lives. Its by-products are all around us and have helped us to gain a greater understanding of the world and to navigate our way around it. We have GPS technology on our phones and in our cars. Every day, new applications for smartphones, tablets and laptops are released that make use of it. Our maps are made with the technology built-in and, as a result of GPS, they are becoming more accurate and richer in the provision of information. Anywhere on earth where there is a clear line to four satellites, there can be GPS.

The background

The origins of GPS can be traced back to the US military during the latter stages of World War II. Roger L. Easton, a scientist working at the Naval Research Laboratory (NRL), had joined the war effort in 1943. He and other scientists were looking at ways of monitoring military vehicles and thought they might be best tracked by spacecraft in orbit. Space flight was still a few years away, but after the war Easton continued his work and laid the foundation for GPS.

In 1955, two years before the Soviet Union launched Sputnik 1, Easton published a paper called 'A Scientific Satellite Program'. The paper spelled out NRL's plans to create a system of satellites around the earth, transmitting signals for scientific and military use. Three years later, the programme began construction and was fully operational by 1966. By the 1970s, a GPS system was coming to fruition whereby a receiver on the surface of the Earth could be found with pin-point accuracy. To achieve this, a signal was sent to and from four satellites with synchronised atomic clocks. However, this breakthrough technology would remain an entirely military endeavour until the 1980s.

TIMATION-I Satellite (rectangular object) shown mounted on the side of its launch vehicle.

In 1983, the Soviet Union accidentally shot down a civilian airliner that had flown off course. Following this avoidable disaster, President Reagan announced that he would allow civilian use of GPS, although the highest grade of use would remain reserved for the military. Commercial bodies largely had to make do with Selective Availability (SA), a far less accurate type of GPS. But in 2000, President Bill Clinton switched off SA and businesses were allowed to tap into the full power of military GPS, which could pin-point objects virtually anywhere on the face of the earth.

In 1983, the Soviet Union accidentally shot down a civilian airliner that had flown off course ... Following this President Reagan announced that he would allow civilian use of GPS.

Commercial impact

Even before the lifting of SA, there was considerable commercial activity involving GPS, including car navigation, recreation, tracking, mapping, equipment manufacture, surveying, aviation and marine activities, along with, of course, military operations. The sheer power that came from the knowledge of knowing where things were, as well as where they were in relation to one another, was too good for businesses to ignore.

Travel companies and logistics businesses benefited hugely from GPS – navigating the world suddenly becomes a lot easier if you know exactly where you are and how far you have to go. GPS made timing, tracking and planning far easier and saved companies considerable time and money.

Big companies working in the energy sector, digging oil wells or creating pipelines, found there were many benefits to knowing exactly where their activities were taking place. GPS made decision-making and planning sessions far easier and gave a bird's-eye view to those in the boardroom of what was taking place in the field.

For large manufacturers and exporters, GPS meant that items could be traced and tracked as they moved around the world. This reduced the threat of theft or accidental loss, minimising misplaced items and lowering insurance costs.

Since the lifting of SA, the commercial potential of GPS, often tied to internet and mobile technologies, has blossomed. There are now scores of devices, services and uses for GPS and it is commonly believed that its full potential is far from fulfilled. Consumers have become the main drivers of GPS technology and more and more applications are being created as a result.

Personal Navigation Devices (PNDs) as stand-alone devices or in phones, are the biggest market for GPS technology; in 2007 it was estimated that PNDs accounted for as much as 90% of the GPS market. Today, most new phones contain some sort of mapping application that harnesses the power of GPS. Car navigation is one of the main consumer uses and devices such as TomToms have proved to be a massive success with drivers. In 2011, the Dutch company had a market capitalisation of over €800m, and it is one of the biggest brands in the sector.

Some companies have developed bespoke devices using GPS technology, which give them a major edge over the competition. eCourier is a great example of a British company that has done this. The business, founded in 2003 by Tom Allason and Jay Bregman, provides courier and delivery services that make the most of GPS. Customers are able to track online in real-time exactly where their delivery is and, crucially for the business, employees are also trackable. The company also benefits from real-time traffic information and even weather patterns. For years, couriers have struggled to expand because the back-office resources required to maintain large numbers of couriers on the road made it difficult for them to be profitable. However, with new smart systems, eCourier looks to have cracked this conundrum.

Some companies have developed bespoke devices using GPS technology ... eCourier is a great example of a British company that has done this.

What happened next?

The commercial impact of GPS looks set to keep escalating. The GPS market is thought to be worth in the region of $75bn worldwide. However, the overall value to business is many times greater than that because it has become a staple technology to so many industries.

Currently, the satellite systems used are the property of the USA but, theoretically, others could take up the mantle in the future. GPS relies on space satellites, and these need to be upgraded as time goes by. This does, however, provide a major commercial motivator for investment in space technology.

Government planners also understand the potential of GPS for far more than just military use. Road-charging schemes are being considered whereby motorists 'pay per mile', a scheme which would have been unthinkable and impossible in the past. For some, such ideas are anathema and open up questions regarding civil liberties and rights.

The inventors of GPS, in particular Roger L. Easton, have been decorated for their exemplary work. Easton himself received a medal and was recognised by the National Inventors Hall of Fame.

30

Fibre optics

When: 1977

Where: UK

Why: The technology can carry vast amounts of information and is transforming communications

How: Freeing glass fibre from impurities gave it the capacity to carry large-volumes of data at high speeds

Who: Sir Charles Kao and George Hockham

Fact: Today more than 80% of the world's long-distance telecoms traffic is carried over optical fibre cables

Communications technology has evolved quickly and it's hard to believe that 20 years ago document transmission was done mainly by post or fax. Yet the standardisation of computer operating systems and the migration of desktop to mobile technology have been adopted so efficiently that global communications now depend on these.

Most businesses can't imagine what they did without them. Add to that the proliferation of video and music sharing on the internet and the exponential growth of social media. The revolution is one that could not have happened without fibre optics to carry the enhanced traffic.

The background

Copper cable, in a variety of forms, has been used for decades to transmit analogue and digital data from point to point. However, copper-based cables have a number of intrinsic limitations. Chief among these is the phenomenon known as signal attenuation – where the signal becomes weaker over distance. The thinner the copper cable, the greater the attenuation. But the thicker the cable, the greater the cost; and with the price of copper at between $9,000 and $10,000 a tonne and rising, that's a major disadvantage. Furthermore, in copper conductors the signal loss is dependent on frequency – which means that the higher the data rate used, the higher the loss.

Another problem is that copper cables are prone to electromagnetic interference when they are laid close to power supply cables or come close to electrical equipment. This can create a high risk of error and corruption of data. Further, signals carried by the cable radiate from its entire length, making it very easy to eavesdrop on what is being transmitted. And copper cables are affected by differences in voltage at either end, providing a source of data errors. Sudden surges, caused by electrical storms for example, can do the same thing, or even melt the cable.

In the 1960s one of the leading research facilities in the field of telecommunications was STL (Standard Telecommunication Laboratories), which had been set up by Standard Telephones and Cables at the end of World War II. Its original location was on Hendon Aerodrome in north London, but in 1959 STL moved, with 500 of its staff, from Enfield to a modern laboratory complex on the edge of Harlow New Town.

In the early 1960s Charles K. Kao started to work at STL while studying for his PhD at University College London. Kao was born at Shanghai in 1933 and had moved to the UK from Hong Kong to take a degree course in electrical engineering at Woolwich Polytechnic; however, it was at STL that he and his associates completed their pioneering work into the use of fibre optics as a viable telecommunications medium.

A young Charles Kao doing an early experiment on optical fibre at the Standard Telecommunications Laboratory in Harlow.

In Victorian times the total internal reflection principle on which fibre optics rely was exploited to illuminate streams of water in public fountains. This principle was later harnessed for short-range optical applications. In the early 1950s Harold Hopkins FRS devised a method of producing a coherent bundle of fibres that he developed into his 'fibroscope', the first device able to transmit a useful image. His work led in 1963 to the invention by Fernando Alves Martins of the fibre-optic endoscope.

Nevertheless silica fibre was still not much use for long-distance communications because of the heavy light loss over a distance. C. K. Kao is credited with the breakthrough that led to modern fibre optics telecommunications, initially by showing that light attenuation within fibres was the result of impurities in the glass that could potentially be removed.

When Kao was appointed head of the electro-optics research group at STL in 1963, he joined forces with a colleague, George Hockham, and together they set about proving that the potential attenuation of light in a glass filament could be 500 times lower than the typical performance of optical fibres at that time.

There was laughter among the audience, so absurd did the idea of light as a transmission medium seem then, even among professional engineers.

The two researchers established that, to transmit optical signals for dependable commercial communications and data transfer, the limit for light attenuation would have to be under 20 decibels per kilometre (dB/km) – however in the early 1960s it was not unusual for optical fibre to lose light at a rate of more than 1,000 dB/km. This conclusion opened the intense race to find low-loss materials and fibres suitable for reaching such criteria.

All that you needed, they concluded, was a purer type of glass. It didn't exist yet, but Kao and Hockham had opened the way for the development of a better medium – if anybody could be persuaded to invest in it. Their study, 'Dielectric-fibre surface waveguides for optical frequencies', was presented in January 1966 at the Institute of Electrical Engineers (IEE). Apparently there was laughter among the audience, so absurd did the idea of light as a transmission medium seem then, even among professional engineers. But Kao and Hockham were to have the last laugh.

At first the industry, too, was reluctant to believe that high-purity glass fibres could be used for long-distance information transfer and could replace copper wires. Kao had to tout his ideas around the USA and Japan over the next seven years before the industry started to sit up and take notice. But eventually a viable medium was developed, a glass that contained fewer than 10 impurities per billion atoms.

In 1966 the American corporation Corning brought together three scientists to create a telecommunications product that would meet the demand for increased bandwidth. Drs Robert Maurer, Donald Keck and Peter Schultz were given a task that was seemingly impossible as defined by Kao: to design a single-mode fibre (100 micron diameter with 0.75 micron core) having a total signal loss of less than 20 dB/km (decibels per kilometre).

But, after four years of experimentation and testing, Maurer, Keck and Schultz developed the first low-loss optical fibre, with an attenuation of less than 17 dB/km. By 1977 they had brought this distortion down to just five decibels per kilometre.

The first field trials were held in June 1977 in the vicinity of STL. BT successfully tested a 140 megabits-per-second (Mbit/s) optical fibre link between Hitchin and Stevenage in the UK. Because the engineers ran real public service telephone calls though the fibre, this date is marked as the birth of fibre optics.

Commercial impact

Traditional phone lines can carry a very limited number of phone connections at a time, for a limited distance. Using the best receiving and transmitting technology, a single fibre nine micrometres in diameter can carry up to 400 billion bits per second, the equivalent of five million telephone conversations.

The other disadvantages of copper are also avoided. Electromagnetic interference is not experienced: fibre optic cable can be laid next to power distribution cables and suffer no interference at all, and it is impossible to eavesdrop on what is being transmitted. Fibre optic cables are unaffected by electrical storms or ground conductivity.

A single [optical] fibre nine micrometres in diameter can carry up to 400 billion bits per second, the equivalent of five million telephone conversations.

Kao predicted in 1983, five years ahead of when a trans-oceanic fibre-optic cable first became serviceable, that the world's seas would one day be littered with fibre optics. Over the last few years fibre optic technology has advanced at a tremendous rate, in a rather quiet and reserved manner, driven by the need for higher bandwidths on long-distance backbone links, as they are called. These include a global network of sub-sea cables such as FLAG (Fibre-Optic Link Around the Globe), a 28,000km-long submarine communications cable containing optical fibre that connects the UK, Japan and many places in between.

While it is difficult to place a commercial value on the use of fibre optics, BCC Research estimated in January 2011 that the fibre optic connector (FOC) market was worth $1.9bn in 2010 and by 2016 will be worth close to $3.1bn.

One of the latest ventures in the field is the network being rolled out by Seacom, the first to provide broadband to countries in East Africa, which previously relied entirely on expensive and slower satellite connections. South Africa, Madagascar, Mozambique, Tanzania and Kenya are interconnected via a protected ring structure on the continent. A second express fibre pair connects South Africa to Kenya and there are connections to France and India.

To developing countries that lack a long-established telecommunications infrastructure, the advent of fibre optics represents an opportunity to develop IT services and support industries; Kenya is rapidly taking over some of the volume of call-centre traffic previously captured by India.

What happened next?

Optical fibre quickly replaced cable in long-distance applications, whether on national POTS (plain old telephone service) or global networks or for secure corporate communication between sites. It is favoured by cable television companies because its high bandwidth capacity makes it ideal for high-definition television (HDTV).

The one remaining problem is the 'final mile' connecting local exchanges to the consumer. The technology is there in different forms, notably Ethernet Passive Optical Network (EPON), but its implementation at a local level is inhibited by the large investment involved. As people demand streamed entertainment and business relies more on global data transfer and video conferencing, the solution to this problem is only a matter of time away.

On 6 October 2009, C. K. Kao was awarded the Nobel Prize in Physics for his work on the transmission of light in optical fibres and for fibre communication. He was knighted in 2010.

31

The electronic spreadsheet

When: 1978

Where: USA

Why: The invention eliminated tedious manual processes and streamlined the business world

How: The idea for the electronic spreadsheet came to Dan Bricklin while he was a student on Harvard Business School's MBA programme

Who: Dan Bricklin and Bob Frankston

Fact: In 1979, software programs that were based on mathematical algorithms were not considered a suitable patent candidate: the first electronic spreadsheet was never patented

I t is now hard to imagine a business environment without spreadsheets. These electronic tables are used for manipulating information such as market research and for planning projects, as well as for the financial forecasting they were originally designed to improve. It is no exaggeration to say that finance departments at the vast majority of the world's businesses and institutions would be unable to do the work they currently do without spreadsheet software.

Yet, in 1978 there were no spreadsheet software packages. VisiCalc, the first spreadsheet for personal computers, helped to revolutionise the day-to-day workings of businesses large and small. In fact it is credited with helping to drive the success of the Apple II computer and introducing the business world to the potential of computers.

The background

In 1978 Dan Bricklin was studying for his MBA at Harvard's Business School. Dan had graduated in computer science from MIT, and prior to going to Harvard had worked for Digital Equipment Corporation, so he was well versed in the potential of computers. Bricklin's Eureka moment came when one of his lecturers was forced to correct values in a financial model manually, after making a small error on the theatre blackboard. Bricklin was struck by how much work was involved in correcting even one small error in a traditional financial model. He paused to dream of a calculator that had a ball in its back, 'like a mouse', and a big display that would allow him to play around with the sums.

Perhaps not surprisingly in such a business-focused environment and at a time when tech companies were big news in the USA, Bricklin decided to test his idea: he would create a real product to sell after graduation. He went on to develop a basic spreadsheet with programmer friend Bob Frankston. This program enabled users to set up a basic table with an interactive user interface akin to word processors; it allowed users to add formulae and featured automatic recalculation, thereby solving the problem Bricklin had observed in his Harvard lecture.

One of [Bricklin's] lecturers was forced to correct values in a financial model manually, after making a small error on the theatre blackboard. Bricklin was struck by how much work was involved in correcting even one small error in a traditional financial model.

Bricklin used the early version of his new program for one of his own Harvard Business School assignments, for which he had to analyse Pepsi-Cola. Using his prototype software, he was able to develop an impressive presentation involving five-year financial projections, dealing with many variables and multiple possibilities: a tough ask in the days of paper ledgers and touch-key calculators, his software saved him countless hours of work.

Bricklin and Frankston developed the prototype further and created a version that was good enough to sell. They named the program VisiCalc.

Commercial impact

The potential of VisiCalc was quickly realised. Dan Fylstra, who had graduated from Harvard Business School around the same time as Frankston and Bricklin, and ran a software company called Personal Software, offered to publish their software. He offered to pay them around 40% of his revenue for sales to individuals and resellers, and 50% for any larger-volume sales to computer manufacturers. According to Bricklin, these figures were based on an initial price for the product equivalent to the Texas Instruments (TI) calculator, less costs, and a split percentage of the profit. Soon after, Bricklin and Frankston formed a separate company under which to do business: Software Arts, Inc was incorporated early in 1979.

Spreadsheets have changed the way businesses operate.

VisiCalc first received public attention at the West Coast Computer Fair in San Francisco, and soon after was launched at the National Computer Conference in New York City, in the summer of 1979. It was not an immediate success, but Bricklin and Frankston were able to use some of the royalty payments from Personal Software to finish converting their product for other computers. VisiCalc was then released to run on the Tandy TRS80, the Commodore Pet and Atari's 8-bit computers. It rapidly became a best-seller, going on to sell 600,000 copies at $100 each.

Neither Personal Software nor Software Arts, Inc made any application for a patent on the software. And while copyright and trademark protection were used and pursued, no further protections were used to prevent others copying the work. Unfortunately, what Bricklin calls the 'enormous importance and value of the spreadsheet' did not become apparent for at least two years. By this stage it was too late to file for patent protection. In any case, in the late 1970s, software programs that were based on mathematical algorithms were not considered to be candidates for patents.

Sales of spreadsheets grew exponentially. But as with any successful product, competitors sprouted up. Although VisiCalc was converted for IBM's PC, Mitch Kapor's Cambridge, Massachusetts-based Lotus Corporation released 1-2-3 for the IBM PC in 1982. This was faster and more powerful and flexible than VisiCalc, and outsold it significantly from then on. Just as VisiCalc had driven sales of the Apple II, so 1-2-3 now drove sales of the IBM PC and helped to establish it as the dominant desktop computer for business. Lotus ended up buying Software Arts in 1985. Microsoft launched Excel, a WYSIWIG ('what you see is what you get') spreadsheet for Apple's Macintosh computer in 1985, which made the power of Lotus 1-2-3 much more accessible to people. Excel was launched for Windows in 1987, and soon became the market leader. As part of Microsoft's Office suite of applications, more than 80% of the world's enterprises are now believed to use Excel. Many more will use other spreadsheets, such as that provided within Sun Microsystems' Open Office.

Just as VisiCalc had driven sales of the Apple II, so 1-2-3 now drove sales of the IBM PC and helped to establish it as the dominant desktop computer for business.

These days, thanks to the power of electronic spreadsheets, five-year financial projections are commonplace. And the use of spreadsheets has grown dramatically, far beyond the finance community. People today use spreadsheets to manage projects and to analyse marketing information and even sports statistics.

Business could simply not be done the way it is today without electronic spreadsheets. The idea of having to tot up all your figures on a calculator, noting each figure down by hand, is just unthinkable.

What happened next?

The success of the electronic spreadsheet is now firmly established; but what of its originators? What of 'father of the spreadsheet', Bricklin, and developer, Frankston? Well, Frankston has become outspoken on modern IT issues, with a focus on internet and telecommunication matters. He is calling for the role of telecommunications companies in the evolution of the internet to be reduced, that of broadband and mobile communications companies in particular.

Bricklin is also active within the industry. He was given a Grace Murray Hopper Award in 1981 for VisiCalc, shortly before selling the rights of the program to Lotus, and is currently president of Software Garden, Inc, a consultancy and software development company. Bricklin has not left his first love behind either: he has developed a collaborative, web-based spreadsheet, called WikiCalc. As such, WikiCalc is joining the new generation of online spreadsheets that have begun to emerge in recent years. The main advantage of online applications is in their multi-user collaborative features. Some online spreadsheet applications even offer stock prices, currency exchange rates and other real-time updates. It seems clear that we will continue to find new uses for these amazing, flexible tools.

32

The Walkman

When: 1979

Where: Japan

Why: The Walkman was the first pocket-sized, low-cost stereo music player and the forerunner for the many innovations that have followed, such as the iPod

How: The need for a portable, personal, stereo audio cassette player was perceived by Sony Corporation. However, the brand was later forced to concede that an independent inventor was first to patent the product

Who: Sony's Akio Morita and engineer Nobutoshi Kihara (and German-Brazilian inventor, Andreas Pavel)

Fact: An estimated 400 million Walkman units have been sold since its launch in 1979

The Walkman brought together technical innovation and smart marketing at just the moment when popular culture needed it. The first Walkman (variously branded as the Stowaway, the Soundabout and the Freestyle before the current name stuck) featured a cassette player and the world's first lightweight headphones. Apparently fearful that consumers would consider the Walkman antisocial, Sony built the first units with two headphone jacks so you could share music with a friend. The company later dropped this feature. Now, more than 30 years and an estimated 400 million units later, it's not in the least unusual for people to be seen walking along the street with headphones on.

The Walkman's universality over the last two decades of the 20th century was supported by outstanding advertising campaigns that targeted a young audience with the concepts of mobility, togetherness and the freedom provided by music.

The background

During the 1970s the audio industry was revelling in the success of the growing home stereo market. The arrival of the transistor, making a portable AM band receiver possible, had created a boom in pocket radios in the 1960s. The 'tranny' became ubiquitous among a generation for whom the Beatles and the Rolling Stones were essential companions. You could take your tranny to the beach, into the woods or on the bus and tune it to illegal radio stations that played great music that affronted your parents and their generation.

Battery-powered one-piece stereo systems grew in popularity during the late 1960s, the sound emanating through two or more loudspeakers. Then boom boxes and ghetto blasters became popular, and supported the capacity of the young to make a statement without being confined to sitting near a home stereo system – which could only impact on the immediate neighbourhood.

Pocket-sized micro- and mini-cassette players were also successfully sold by companies such as Panasonic, Toshiba and Olympus. But Sony was already the market leader in making recorded music accessible.

In 1949 the Tokyo Telecommunications Engineering Corporation (TTEC), as the company was known then, had developed magnetic recording tape and in 1950 it went on to develop and market the first tape recorder in Japan. It produced the world's first fully transistorised pocket radio in 1957. While the transistor was developed by Bell Labs and produced by Western Electric, it was Sony that first used it for a small pocket radio, creating a new market in the process and giving it the opportunity to claim the platform that started a worldwide youth craze. The radio's success led to more firsts in transistorised products, such as an eight-inch television and a videotape recorder.

But with a name like TTEC, the brand was never going to be cool, let alone become a global brand. In 1958 its founders, Akio Morita and Masaru Ibuka, had the foresight to change the name of the company to Sony. Based on the Latin word for sound, 'Sony' was also a name that would resonate with Japanese youth culture. Sonny-boy was the street name for a whiz kid at that time: Morita and Ibuka wanted to create a young and energetic image for their company.

Morita can further claim to have been the inspiration behind the development of the Walkman. As well as closely identifying with global hip culture, he was very keen on art and music, and was also something of an opera buff, and he wanted something that would enable him to listen to his favourite songs on the long-haul flights he frequently took. For that it would have to reproduce music every bit as well as a good-quality car stereo, yet remain portable, allowing the user to listen while doing something else. He observed how his children and their friends seemed to want to play their music day and night, and how people were increasingly taking bulky stereos to the beach and the park.

Engineer Nobutoshi Kihara put together the original Walkman device in 1978. Early in 1979 Morita convened a meeting in which he held up a prototype derived from Sony's Pressman portable tape recorder. He gave the engineering team less than four months to produce the model.

'This is the product that will satisfy those young people who want to listen to music all day' ... [said Morita.]

The engineering department was initially sceptical, since the device had no recording capacity. All the technology was pre-existing, and in many ways the Walkman can be considered to have evolved from cassette recorders developed for secretarial use and journalists, estate agents, doctors, lawyers and other professionals. Morita was adamant that recording their own audio was not something his target customers would want to do. 'This is the product that will satisfy those young people who want to listen to music all day. They'll take it everywhere with them, and they won't care about record functions. If we put a playback-only headphone stereo like this on the market, it'll be a hit,' he said.

Internal doubts were not silenced. As development neared the final stage the production manager, Kozo Ohsone, was concerned that if Walkman failed it could take his reputation with it. While he had already spent over $100,000 on the equipment to produce the injection-moulded cases, Ohsone did not want to risk over-production. So though he had been asked to produce 60,000 units, he decided to produce half that number. If sales took off, 30,000 more could be manufactured in short order.

The original metal-cased silver and blue Walkman TPS-L2, the world's first low-cost portable stereo, was launched in Tokyo on 22 June 1979. Before the release of the Walkman, Sony seeded the market by distributing 100 cassette-players to influential individuals, such as magazine editors and musicians. Journalists were treated to an unusual press conference. They were taken to the Yoyogi Park in Tokyo and given a Walkman to wear.

The journalists listened to an explanation of the Walkman in stereo, while Sony staff members carried out product demonstrations. The tape invited the journalists to watch young people listening to a Walkman while bike riding, roller-skating and generally having a good time on the move.

The original Sony metal cased silver and blue Walkman TPS-L2.

Despite this, to the alarm of management, in the first month after the launch a mere 3,000 units were sold. However, within another month Walkman sales started to take off and the original run of 30,000 had sold out – Ohsone had to ramp up production to meet the demand from local consumers and tourists.

The following year, Walkman was introduced to the American market, and a model with two mini headphone jacks, permitting two people to listen at the same time, was marketed in the UK and other territories. At first, different names were used in different markets – Soundabout in the USA, Stowaway in the UK and Freestyle in Sweden. Morita initially resisted the name Walkman, though it was put forward for the same reasons the company had called itself Sony: it incorporated the idea of movement and neatly echoed Superman. On a business trip, Morita found that both the French and the people he met in the UK asked him when they would be able to get a 'Walkman'. The name had preceded the product and, after the lightweight Walkman II was brought out in 1981, the alternative titles were dropped.

Miniaturisation followed, with the creation in 1983 of a Walkman roughly the size of a cassette case. The WM-20 featured one AA battery, head, pinch roller, and headphone jack arranged in a row, with the cassette placed horizontally alongside this structure. From this point forward, the Walkman was established as a modern necessity, easily carried in a handbag or pocket. In 1984 the Walkman Professional WM-D6C was introduced, with audio quality comparable to the best professional equipment. Thus Sony led the market, despite fierce competition, throughout the 1980s and most of the 1990s.

Commercial impact

The launch of the Walkman in 1979 changed forever the way people listened to music, because for the first time it was possible for people to travel with their favourite records and not disturb anybody else by playing them. In doing so, it kick-started an entire industry and prepared the market for the digitisation of music. When production of the cassette-based Walkman ceased in 2010 after 31 years it boasted incredible lifetime sales of more than 220 million. In all, Sony claims to have sold more than 400 million units bearing the Walkman name and for now continues to sell CD and mini-disc versions of the product.

Since the first shipment of the Walkman, almost every other company in the market has made its own portable cassette player. Nevertheless the Walkman's effect on the market was so huge that even these competing models became referred to as 'Walkmans' in much the same way that vacuum cleaners are called 'Hoovers', regardless of their brand. The other names marketed – Toshiba's Walky, Aiwa's CassetteBoy and Panasonic's MiJockey – were nothing like as memorable.

The launch of the Walkman in 1979 changed for ever the way people listened to music [...] it kick-started an entire industry and prepared the market for the digitisation of music.

The Walkman was the product that put the name of Sony on the lips of an international clientele. An imaginative programme of advertising and marketing helped to convey the message that to be young, active, sporty, vigorous and cool, not to mention rebellious, you had to be seen with a Walkman. The success of Sony's subsequent products, including the PlayStation and its digital camera range, have been boosted by this powerful association.

Harder to prove, but eminently arguable, is the case that the Walkman played its part as an ambassador for Japanese exports. The Walkman helped to overcome the prejudice that Japanese goods, including cars, were somehow imitations of western prototypes.

However, the brand also attached itself to psychology and sociology. 'The Walkman Effect' has been defined by some researchers as promoting isolation, self-absorption and even narcissism, while Sony has been at pains to argue the opposite, describing it as 'the autonomy of the walking self'. Sony claims that the device 'provides listeners with a personal soundtrack to their lives', allowing its users to make even the dullest activities interesting and adding a bit of panache to their routines. As MP3 and mobile phone technology has proliferated, with Bluetooth headsets and earpieces permitting people to connect without wires, the debate around isolation has intensified – yet it is still defined as 'The Walkman Effect'.

What happened next?

However innovative the Walkman concept as developed by Morita and Kihara – and notwithstanding that it was put together from intellectual property that resided within the corporation – the design was the subject of disputes for many years and Sony eventually had to concede that its innovative design was not, in fact, the first of its kind. A German-Brazilian inventor called Andreas Pavel had produced a portable stereo cassette player as early as 1972. In 1977 Pavel patented his Stereobelt in Italy, Germany, the USA, the UK – and Japan. Just a year after the Walkman was launched, in 1980, lawyers for Pavel and Sony started talks. As a result, Sony offered to pay Pavel limited royalties, covering the sale of certain models – but only in Germany.

This was not acceptable to Pavel, who wanted to be recognised as the inventor of the Walkman and to be paid royalties on sales in all the jurisdictions covered by his patents. Akio Morita, who had a strongly proprietorial interest in his company's most successful product, was never going to agree to that, and Pavel pursued lawsuits that left him out of pocket to the tune of $3.6m and close to bankruptcy. He would not abandon his claim, however, and in 2003, after more than 20 years of court battles, Andreas Pavel was given several million dollars in an out-of-court settlement and was finally accepted as the Walkman's inventor.

However innovative the Walkman concept [...] the design was the subject of disputes for many years and Sony eventually had to concede that its innovative design was not, in fact, the first of its kind.

The cassette Walkman continued to be produced until 2010 – in fact at the time of writing Sony is still manufacturing one model in China – but from the mid-1990s CDs started to replace cassettes. Manufacturers (including Sony, which produced its first Discman in 1984) migrated to personal disc players and, dealing a blow to car stereo manufacturers, phased out cassettes in favour of discs. The final nail was driven into the cassette's coffin when Apple's iPod and a host of MP3 players and mobile phones provided consumers with the digital storage capabilities to carry a lifetime of music on their person at all times, without having to carry cassettes or discs.

With the cassette all but dead, Sony has worked hard to extend the Walkman brand to a series of subsequent digital music platforms. This started with the Discman (later CD Walkman), then the MiniDisc Walkman, and a number of digital music players aimed at capturing the market opened up by the iPod. More recently, Sony released the Walkman X Series – a touchscreen audio and video player.

Then, on 1 March 2005, Sony Ericsson, a joint-venture company established in 2001, introduced the W800i – the first of a series of Walkman phones, capable of 30 hours of music playback. During the following year it sold three million Walkman handsets, showing that the brand had not lost its magic, despite being eternally associated in the public imagination with Sony's original silver and blue cassette player.

1980s

33
The personal computer (PC)

When: 1980

Where: USA

Why: The PC revolutionised the way businesses operate and created a major product category

How: IBM's iconic PC gave desktop computers credibility with business

Who: IBM's Don Estridge, his team ... and Mr Bill Gates

Fact: *Time* magazine named the computer 'Machine of the Year' just months after the PC's 1981 launch

Mighty mainframe computers had been around since the early 1950s. These machines took up whole rooms and required specialists to operate and maintain them. They were expensive, and only used by very large corporations and government institutions.

The 'IBM compatible' PC, as it became known, succeeded to an unprecedented degree and remains the tool used by almost every business around the world, as well as the device most of us have at home to browse the web, store pictures and music, send and receive email, write documents and so on.

The background

The mid-1970s saw the first consumer computers appear on the market. But these early machines were a different breed to the Windows-fired powerhouses that sit on countless office desks today. So-called 'micro-computer' project kits were aimed primarily at subscribers to such niche publications as *Radio-Electronics* and *BYTE* magazine: enthusiasts who were interested and skilled enough to build the machines at home. Such early-comers included the SCELBI-8H and the Mark-8 Altair. These machines had to be programmed by the user, and often came with just 256 bytes of Random Access Memory (RAM), with indicator lights and switches serving as the only input/output devices – there were no mice, keyboards or even screen displays.

It was not until the late 1970s that technology had improved enough to make the machines worthy of the tag 'home computer'. The late Steve Jobs and business partner Steve Wozniak launched the Apple I computer in 1976; their second computer, the Apple II, as well as Commodore's PET and Tandy's TRS-80 formed the first generation of microcomputers that were designed and affordable for home or small business use. And they had keyboards and 'visual display units' (VDUs) so that people could see what they were doing.

These micro-computers had enough processing power and memory to support the basic programming, word processing and gaming requirements of home users. A cottage industry grew up supplying software and accessories for these computers – games, programming languages, very simple word processors and databases, printers, disk drives and the like. Small businesses began to buy these computers for use in the workplace, though the majority of them were used at home. The computers were big news, as were the entrepreneurs behind the companies making them – which experienced phenomenal growth and profits.

IBM was the global leader in computing at this point, selling almost exclusively to very large companies and governments. It saw the new trend for small microcomputers, and the potential for businesses to use them, and naturally wanted to participate in this new market.

So it came up with a plan to launch its own PC. IBM was well known for doing everything itself – it would not use other suppliers' parts, it would normally create them all itself. However, it realised that this new market was growing so fast that if it followed its normal approach, it would take so long that it might miss out on much of the new business. So it decided to buy in various components for its new PC, little realising at the time just how big the ramifications would prove to be.

Someone suggested that they contact a small company called Microsoft, and so they met Bill Gates, who assured them that he could deliver a suitable operating system on time.

IBM established 'Project Chess': a team of 12, headed by one Don Estridge, which was put together to bypass usual company procedures so the new computer could be developed in a very short period of time. In the end, it took the team just 12 months.

During development, Estridge and his team vacillated over which operating system to use in the new machine. They tried to license the rights to one of the microcomputer operating systems that prevailed at the time, but couldn't. Someone suggested that they contact a small company called Microsoft, and so they met Bill Gates, who assured them that he could deliver a suitable operating system on time. So Microsoft won the contract. In one of the world's best-ever business deals, Gates then bought a similar operating system from an unsuspecting small business, Seattle Computer Products, for a modest one-time payment, and adapted it for IBM.

IBM launched its PC in 1981, to much fanfare. And it worked. Businesses large and small bought the computers on the back of IBM's name. Software developers developed programs for the new computer in the expectation that IBM's new launch would sell well, fuelling demand. Lotus 1-2-3, in particular, drove substantial sales of the PC, as you can read about in the chapter on spreadsheets in this book.

Don Estridge, leader of 'Project Chess' led the group to develop IBM's new computer in just 12 months.

181

Commercial impact

The launch of the desktop computer is, arguably, the single most important business development of the last 50 years. It has led to many other developments, such as the widespread use of the internet and email, but even before that it revolutionised the workplace – how people did their jobs, and also what jobs they had to do. The PC was powerful enough, and backed with sufficient software, to automate all sorts of administrative functions that had previously had to be done by hand.

The IBM Personal Computer of 1981.

And the PC sector exploded into a fast-growing, highly profitable market, with hardware and software manufacturers and resellers, and book and magazine publishers and event organisers all springing up to deliver what businesses globally wanted from this wonderful new technology.

The launch of the desktop computer is, arguably, the single most important business development of the last 50 years.

Of course, some companies benefited far more than others. IBM's decision to outsource its development work had many repercussions. IBM's use of bought-in components for its PC meant that it was unable to stop other manufacturers buying a similar set of components themselves and making careful copies of the PC that did not infringe intellectual property rights. A year after IBM's release of the PC in 1981, the first IBM-compatible computer was released by Columbia Data Products. The Compaq Portable followed soon after, and the gates were open: Dell, Compaq and HP and others were able to manufacture PCs that operated like an IBM machine. And the company that really enabled this to happen was Microsoft, which licensed its operating system to many manufacturers. The term 'IBM compatible' was widely used for PCs that would run the same software and accessories as IBM's PC.

In 1984, Apple launched its Macintosh computer, which used a mouse and the sort of interface we are all used to today – making its computer far easier to use than IBM's. However, Apple was the only company to supply these computers, which it priced higher than most IBM-compatible PCs; and in a repeat of the video-recorder wars a decade earlier, the better technology lost out to the cheaper computer with more software available for it. IBM, Compaq, Dell and myriad others made vast profits from selling their personal computers; Microsoft became a global giant on the back of sales of its operating systems and subsequent Office tools. Intel, too, grew substantially on the back of supplying almost all the processors at the heart of every PC. And Apple, at one time the leading personal computer manufacturer with its Apple II, dwindled to a small, niche player, though much loved by its loyal followers, often in the creative and education sectors.

What happened next?

The price of PCs fell amid strong competition, and technological advances enhanced the PC's power and capabilities every year, helping it to establish a legacy almost unprecedented in modern times.

PCs are now a fixture of modern life. Over 500 million were in use worldwide by 2002, and more than two billion personal computers are thought to have been sold since the earliest versions hit the market. These days, PCs are as much a part of home life as they are of the office landscape: at least half of all the households in Western Europe have one. And the number continues to rise; a billion personal computers were believed to be in use in 2008, and this is expected to double by 2014.

Presently, PCs in all their forms – desktop computers, laptops and servers – are the world's main computing platform. They are used for any and every purpose, from paying bills and accounts, to chatting and playing games.

These days, PCs are as much a part of home life as they are of the office landscape: at least half of all the households in Western Europe have one.

It may not always be this way, with smartphones and tablet computers increasingly performing functions that previously had to be carried out on a PC. By the end of 2011, it is predicted that there will be more people buying smartphones and tablet computers than PCs, with the installed user base sure to surpass that of PCs shortly afterwards. Today it is impossible to say for sure the extent to which computing will move on to these mobile devices, but it is certain that the trend will be substantially towards mobile. It all began with the simple PC though.

34
Infrared remote controls

When: 1980

Where: Canada

Why: The infrared remote control has changed forever the way people watch television and asserted family hierarchy in the home

How: A cable box engineer created an accompanying television remote control, which used infrared light technology

Who: Paul Hrivnak

Fact: The first wireless remote control had to be abandoned after its photosensitive receiver cell was activated by beams of sunlight

35

The Post-it note

When: 1980

Where: USA

Why: The humble Post-it note has become a staple of communication in the western world

How: The 'failure' of a new adhesive prompted experimentation, which led to the Eureka moment

Who: 3M's Spencer Silver and Art Fry

Fact: By 1999, Post-it notes were generating sales of more than $1bn for 3M

The ubiquitous adhesive yellow pads of paper have become such an office staple that it's hard to believe they've only been around for three decades. Post-it notes feature prominently in almost every office store cupboard and are used by everyone, from CEOs to data-entry clerks. The humble Post-it note introduced a brand new method of communication into the US office, after going on sale in 1980. A fantastic example of a product that boasts versatility and flawless simplicity in equal measure, it offers a blank canvas for the office, or even domestic, to-do list and remains the perfect explanatory accompaniment to the memo left on a colleague's desk.

The background

The Post-it note had a rather long gestation period at 3M, the multinational conglomerate behind the product. It was back in the late 1960s that chemist and 3M employee, Dr Spencer Silver, was working on a range of new adhesives. One of the substances he developed was a new but deeply flawed sticky substance. Silver's adhesive was unable to achieve a complete bond with the surfaces it came into contact with, essentially rendering it useless for any kind of permanent sticking. However, instead of discarding the substance, Silver set about trying to find a use for such a mild adhesive. Trialled uses for the substance included a product called the Post-it Bulletin Board – a simple notice board covered with a layer of Silver's adhesive which could house memos and flyers without the need for pins.

Although the product was launched, sales never really took off, but it did keep Silver tinkering with his substance and, more importantly, gave him cause to discuss it with his colleagues. One such colleague was Art Fry, who took an immediate interest after seeing one of Silver's product presentations. Chemical engineering graduate Fry had started out as a salesman before working his way up the ranks at 3M, eventually landing a role within the product development division. He instantly saw the potential in Silver's

Art Fry with his simply but effective invention – the Post-it note.

adhesive and set about thinking up better uses for it. The Eureka moment came during a church service. Fry, a member of the choir, was constantly looking for better methods of bookmarking hymns. The tiny pieces of paper he used were forever falling out of his hymn book, but with a small amount of Silver's adhesive, the strips of paper stayed in place, thus forming the perfect bookmark.

3M's policy of allowing staff a percentage of their working hours to tinker with their own projects meant Fry had time to start developing his bookmarks idea. He adjusted the adhesive's properties so that it left no residue on book pages and started to hand out prototypes to colleagues for feedback. The trouble was, the bookmarks proved too durable. Everyone he gave them to just used the same one over and over, meaning there was no need to consume more. It wasn't until he used one of his bookmarks to annotate a report he was reading that he discovered the full-potential of using Silver's adhesive on strips of paper.

Although the product was launched, sales never really took off, but it did keep Silver tinkering with his substance and, more importantly, gave him cause to discuss it with his colleagues.

Between them, Fry and Silver had come up with a mobile notice board, and an entirely new office communication method. Fry started handing out stacks of the sticky paper sheets to colleagues and very soon, the whole company was using them, and 3M decided to give Fry the resources he needed to develop the product commercially.

Commercial impact

Nearly a decade after Silver had first developed the basis for the Post-it note adhesive, 3M started test-marketing Fry's invention under the name 'Press & Peel Pads' in four cities – Denver, Richmond, Tampa and Tulsa. Reception was muted, to say the least. But both Fry and 3M were reluctant to give up on the product, given that it had been so popular with the company's own staff. In a last-ditch attempt to generate a buzz around Press & Peel Pads, 3M launched a more focused and resource-intensive campaign in Boise, Idaho. Samples were handed out to offices, stationery stores were persuaded to put up point-of-sale displays, local newspapers were convinced to run stories on the product and, most importantly, 3M sales temps were sent out to do demonstrations.

The extra marketing resource proved worthwhile, and the city responded with great enthusiasm, giving 3M the confidence to commit to a full commercial launch in 1980.

The initial product was launched in two sizes – 3 x 5 inches and 1.5 x 2 inches, with a price tag of just under a dollar for each 100-sheet pad. 3M management was still unhappy with the name, however, and it was at this point that the product name was changed to Post-it notes, in an attempt to tie them in with the Post-it Bulletin Board. It was thought that aligning the two products in terms of branding would create better consumer awareness of both.

Once the full weight of the 3M commercial team was behind the product it became profitable within a year, in spite of the massive quantity of free samples being given away. By 1984, sales had reached $45m, and 15 years later revenue topped the $1bn mark. The product's patent ended more than a decade ago and several copycat products have since been introduced onto the market, but the product name and famous 'canary yellow', the colour of some scrap paper used during the prototype stage, are still trademarks of 3M.

What happened next?

But what became of the men behind the Post-it? They were never given shares, or formally compensated for the development of the product in any other form than their 3M salary, but they remained with the company and both went on the develop other products, although none was anywhere near as successful.

Today, 3M manufactures more than 4,000 products under the Post-it brand that are used in offices and homes around the globe. As technology has evolved, and digital methods of communication such as email and instant messaging have become part of the everyday fabric of modern life, the Post-it note still has an important role to play. Post-it branded products are still used for indexing, reminder notes and document annotation, despite the onslaught of electronic communication. It's unlikely they'll become redundant while paper still changes hands and memos are still printed. And even if that day does arrive, digital versions of the yellow pads have been adorning PC desktops for years, proving that the principle of the simple Post-it note will linger on for years to come.

Today, 3M manufactures more than 4,000 products under the Post-it brand that are used in offices and homes around the globe.

36

The compact disc (CD)

When: 1981

Where: Germany, Holland and Japan

Why: The storage of data on optical media revolutionised the music and software industries

How: The connection of an optical disc to a player through light

Who: Invented by James Russell, later produced by Sony and Philips

Fact: Over 200 billion CDs were sold in the 25 years after they were launched

Compact discs (CDs) helped to usher in the digital age of the music industry, and for the last part of the 20th century were the medium of choice for the record-buying public. The invention of the CD and the associated technology changed the music industry for ever, as well as many other sectors, including some which are not closely tied to media. CD technology lent itself to DVD and to CD-ROM, and thus moved into offices and homes for uses other than music and entertainment.

Even though music lovers might now prefer an MP3 to a CD, digital lasers and CD technology are still a part of our lives and are generating revenue directly or indirectly for many types of businesses. The creation of the CD was one of the great business ideas of the age and its legacy continues to this day.

The background

James Russell was a renowned inventor and scientist working at Battelle Memorial Institute (BMI) in its new Pacific Northwest Laboratory in Richland, Washington State, when he began the initial research that led to the invention of the CD. It was 1965 and Russell, a music lover himself, was determined to find a better way of playing music than vinyl records. He was frustrated at the way vinyl was prone to scratching and warping, as well as dust interference. Russell yearned for a clearer and more definite sound recording and believed others shared this desire.

One Saturday afternoon he sketched out a system where no contact between the recording and the player would be made ... he thought a light or laser beam could be used.

Russell conducted numerous experiments and soon concluded that the main problem was the scratch made by the stylus needle on the record. He tried all sorts of things to make the sound better, including different types of needle, but all his solutions reached a dead end and the essential problem remained: physical contact between the needle and the media, no matter how slight, would always lead to problems.

One Saturday afternoon he sketched out a system where no contact between the recording and the player would be made. He thought that a light or laser beam could be used to transmit information from the disc to the player. He was aware that digital data recording was already in existence, used in punch cards and magnetic tapes. His reasoning was that binary zeroes and ones could be replaced with dark and light. The information simply needed to be condensed in order to work in the new format. Over the next few years Russell toiled away, and eventually he created the world's first digital-to-optical recording and playback system, which he patented in 1970.

To Russell's mind this was ground-breaking research and the potential was enormous. He saw that this technology was not just confined to music but could also be used to store and play all kinds of data. He saw the commercial potential of his invention and hoped that a manufacturer would take it forward. However, although his invention did excite some interest from magazines and periodical publishers at the time, he found no commercial backers willing to take it on. Undeterred, Russell continued to work on and refine his ideas and built up a large collection of patents for CD technology. He also created prototypes and built up a body of work that was to have a major impact on the optical and digital revolutions that would be unleashed later in the century. Meanwhile, he continued to send out information to technology manufacturers about CDs in the hope they would be interested. His prototypes were viewed by many interested parties at the time, although take-up was still slow. Russell was a man ahead of his time, though thankfully, for him, not too far ahead.

Like many inventors, he was not alone in his pursuit of new sound-recording techniques and digital data. At Philips two engineers, KlassCompaan and Pete Kramer, had created a glass disc that could be read by a laser beam – to some this was in fact the very first CD.

Eli S. Jacobs, a New York-based venture capitalist heard about Russell's inventions and was convinced of their value. He bought the licensing rights from BMI and established a company, the Digital Recording Corporation, to further enhance the product and to find commercial backers. However, Jacobs was not just interested in music but also in video, and his backing led to the creation of a 20-minute video disc in 1973. Jacobs thought CDs could be used for playing films – and he was right, but also a little ahead of his time.

Philips proudly introduce the launch of the CD.

Large electronic manufacturers such as Philips and Sony were now starting to realise the potential of digital recording and playback and set to work on establishing the rights and the technology that would enable them to do this. In 1978, Sony publicly demonstrated a working optical disc with a playing time of over two hours. Shortly after, Philips also demonstrated its technology and the two companies worked together to create the first commercially available CDs.

The companies understood that CDs were going to be highly disruptive to the music market. However, their merits had to be sold to the public. They also faced the problem of adoption, with expensive players necessary for the use of the new format. It would be very costly and risky for either company to go it alone; they understood that if they both brought out similar but separate products, then this would cause confusion among the public and undermine their cause. The two companies would have to work together to launch the products and move the public towards a new form of music playing. Although joint technology ventures such as this are now fairly common, in the late 1970s such an approach to business was rare and the collaboration was something of a leap of faith.

In 1980, the two companies created a 'red book' of standards and definitions which set out how CDs would be made. Then, one year later, at a plant in Germany the first test CDs under the agreed standards were made, and the preparations to go to market were complete. Sony released Billy Joel's 52nd Street on CD and it was sold alongside CD players to entice customers to try the new format. It also embarked upon a memorable advertising campaign, with the slogan 'Perfect Sound Forever' – a reference to the deterioration in fidelity that vinyl records suffer over time.

Commercial impact

The very first album to be recorded for the CD was Visitors by Abba, and by 1983 many other musicians were following in their tracks and releasing their albums digitally. There was much media attention surrounding CDs. This helped to create further excitement among the public and soon CDs and players were being bought by the public in their droves. In 1985, Dire Straits' Brothers in Arms album sold over a million CDs, and by this point it was clear that CDs were going to be a tremendous success.

The Sony CDP 101.

The format soon proved its utility beyond that of being a superior alternative to analogue sound recording, driving another revolution in business with the development of the CD-ROM.

In the years to come more and more music would be released on CD, and soon it was believed that the death of vinyl was inevitable. Record companies began to focus on CDs in the same way they had once focused on vinyl and cassettes, most of them completing migration to the format by around the mid-1990s. By the early 2000s, vinyl records were a rarity and the dominance of the CD in the music world was firmly established.

But the format soon proved its utility beyond that of being a superior alternative to analogue sound recording, driving another revolution in business with the development of the CD-ROM (Read-Only Memory).

The optical technology used in a CD is digital, meaning that the storage of any kind of data is possible. In 1985 Sony and Philips collaborated once again to create 'Yellow Book', a universal standard used for the storage and retrieval of any kind of data on a CD. When CD-ROM drives for computers became affordable by consumers in the early 1990s, the CD transformed computing. The format could store thousands of times more data than could the existing technology (floppy disks and tapes), meaning that large-scale storage of multimedia such as video became viable for the first time.

The CD-ROM soon rendered other formats obsolete for the distribution of software and other data on PCs; an example of the effect the format had on the computing world is in the changing nature of PC games, which began to incorporate high-fidelity music and live-action cutscenes (often gratuitously at first) into the experience, made possible by the increased storage capacity of the CD. Dedicated games consoles began to use the formats, starting with the Sony Playstation in 1995, and soon optical discs were standard in the computing world, relegating older formats to the realm of memory.

By the 1990s the world's major electronics manufacturers were involved in the production of CDs, CD players and CD drives. Now the interest turned to what could be achieved visually as well as aurally. A joint venture of technology firms including Apple, Dell and Sun Microsystems brought forward the standards and definitions of DVD technology. In doing so they were fulfilling what Eli S. Jacobs had foreseen some 20 years before.

CDs as a storage medium for data has now been largely supplanted by newer formats such as DVDs and flash drives, which offer increased capacity, but their influence should not be forgotten; 200 billion CDs have been sold worldwide in the 30 years since Sony and Philips finalised the standard in Germany, and emerging optical technologies such as Blu-Ray owe their entire existence to the small silver disc.

What happened next?

The music CD is also rapidly falling out of favour with consumers. The development of digital music formats, beginning with the MP3, saw instant delivery of music become a reality. Many people are no longer inclined to visit their local record store to buy physical media, preferring the ease and convenience of online MP3 marketplaces such as iTunes; indeed digital has all but killed the singles market, with downloads accounting for over 99% of total sales. Physical album sales remain the traditional domain of the CD, with 2011 figures putting its share at 82.2%, but this is a figure that is in steady decline, with sales falling by 12.9% between 2008 and 2009.

Although CDs have proved to be but a stepping-stone in the unending quest for smaller and more efficient delivery of content, they can take the credit for having turned the music industry on its head.

These falling sales are of little concern to the record companies, who in any case have embraced the rise of digital, as it saves vast amounts of money on the cost of production. The implications for the record shop business, however, are grave; one only needs to look at the demise of former high street fixtures Woolworths and Our Price to know that times are hard for the industry. Although vinyl is making a modest resurgence, with sales of the once-moribund format rising by 55% in the first half of 2011, CDs have been left out of this upswing, perhaps still too new a technology to be seen as collectable and too old to be seen as useful.

Although CDs have proved to be but a stepping-stone in the unending quest for smaller and more efficient delivery of content, they can take the credit for having turned the music industry on its head, as well as being the first of many optical media formats to follow.

37

The 20-70-10 rule

When: 1981

Where: USA

Why: This changed the way many companies manage, motivate and fire their people

How: As the CEO of General Electric, Jack Welch strove for efficiencies in the workplace

Who: Jack Welch

Fact: Welch grew General Electric's revenues from $26.8bn to $130bn

Jack Welch has earned a reputation as one of America's least compromising CEOs. The man who ran General Electric for 20 years has gained admirers and detractors in near equal measure. To some, he is the pioneering business leader who led corporate America through a slump and into unrivalled growth. For others, he is the brutal cost cutter and job slasher who put thousands out of work in the pursuit of profit and self-aggrandisement.

One of Welch's most famous business maxims – the 20-70-10 rule – is just as polarising as the man himself. To put it simply, at the end of each year Welch would fire his least-performing 10% of managers, put 70% into training and education, and offer stock options and rewards to the top 20%.

The background

Welch joined General Electric (GE), one of the world's biggest companies, in 1960 and worked his way towards the top of the company until achieving the role of CEO in 1981 – a post he would retain for the next 20 years. The future could have been very different because, after little more than a year at GE, he actually threatened to leave the business following a less than satisfactory pay rise. Welch felt he was being overlooked by the management of the enormous company. However, a senior colleague persuaded him to stay, on the understanding that an employee as talented as Welch should and would get every opportunity to shine.

Over the course of his career at GE, Welch became increasingly interested in efficiencies, motivational practices and any business strategies that could increase a company's performance. There was, perhaps, no better company than GE to help a budding manager develop. GE was, and still is today, a vast company with direct control and stakes in a wide range of different industries and markets. Some of these were world class, while others were underperforming or unprofitable. This environment provided Welch's keen mind with much to consider – examples of both good practice and bad.

Jack Welch takes over as Chairman at General Electric.

The 20-70-10 rule became the bedrock of Welch's management style: invest in and improve what you do well and scrap anything that is failing.

Welch developed the belief that about 20% of the managers at GE were exceptional and should be rewarded more than their counterparts (these were Group A). However, about 70% were less dynamic but still productive and useful; these were the largest group, Group B. Group B, he thought, should be given training to live up to their potential. Meanwhile, the bottom 10% (Group C) were, in Welch's opinion, largely ineffective and he decided it would be in the interests of all if they were let go. The 20-70-10 rule (also known as differentiation) became the bedrock of Welch's management style: invest in and improve what you do well and scrap anything that is failing.

Commercial impact

After Welch became CEO of General Electric in 1981, he set about cutting and refining the business. Welch had little time for aspects of the company that were not performing well. He wanted to be either No. 1 or No. 2 in the world, and if there was no chance of the company achieving this high status then he would rather let the business go completely. His view on managers was little different, he felt there was no point in investing time and money in an individual who was never going to make the grade.

Welch's differentiation strategy rewarded those who performed well with stock options and other benefits, and during his time as CEO the numbers of employee shareholders at GE rose significantly. While critics tend to aim their fire on his 10% cutting policy, they sometimes neglect this aspect of his approach. Providing stock options to staff inspires them to try harder and perform better because they have a stake in the business.

Meanwhile, for those in the 70% group, the prospect of greater rewards is an incentive, whereas the 10% cull keeps everyone on their toes. However, Welch has since advocated a period of three years of appraisals before any significant reductions in staff numbers are made. He wanted such decisions to be made on genuine management information, not just on gut instinct.

What happened next?

Welch retired from GE in 2001 but has continued to spread his ideas on business and management. He is an author of books such as *Winning* and *Straight from the Gut*. He also lectures to business leaders and students around the world, regularly appearing at MIT's prestigious Sloan School of Management. His work has won him plaudits from many senior managers. The differentiation model has been employed by other large organisations such as Microsoft, Motorola and Dow Chemical and is credited by many company owners as the bedrock of their success.

For some, differentiation is ethically dubious and creates a hyper-competitive atmosphere, rather than an environment where team building can take place.

However, others have not been quite so complimentary. Critics of Welch suggest that his insistence on removing 10% of the managers is both arbitrary and unscientific. Indeed, even Welch would concede that it was never a precise measurement. As time goes on, the bar is increasingly raised, and so those in the 70% are pushed ever further. Smaller companies argue that only businesses as large and with as many resources as GE could implement such a strategy. Also, to carry out a policy such as differentiation you would require a lack of legal restraint, which is not always possible. For some, differentiation is ethically dubious and creates a hyper-competitive atmosphere, rather than an environment where team building can take place.

Welch's differentiation model has now been quietly shelved or modified by many organisations that once practised it, most notably at GE itself. The harshness of the 10% cull has led business leaders to seek more flexible strategies. It is, perhaps, a shame that this headline-grabbing aspect of a model that also embraces team building, rewards and ownership came to dominate so much of the debate. Nevertheless, the differentiation model continues to influence theorists and business leaders and will do for some time to come, even if their own models take on board just parts of Welch's revolutionary idea.

38

The digital camera

When: 1981

Where: Japan

Why: Digital cameras have enabled amateur and professional photographers alike to take almost unlimited images, delete at will, with no need to carry film rolls

How: Storing images on floppy disks allowed digital camera pioneers to dream for the first time of a world without film

Who: Sony

Fact: In 2010, a record 143 million units were sold around the globe as the digital camera approached its 30th birthday

The development of film-free image capture and storage has completely transformed the camera market in just a couple of decades, enabling camera makers to bring high-end features to ordinary users and providing a springboard for camera sensors to find their way into all manner of other devices, from mobile phones to laptops.

It has removed the need for the costly, time-consuming process of developing and printing, liberated photography and made it a far more appealing and affordable prospect than ever before.

Today, digital imaging is a multi-billion-pound industry and even the professional-specification film cameras of years gone by are all but worthless. And with digital technologies evolving and advancing all the time, there's no sign of things slowing down any time soon.

The background

The camera market's move away from film and on to digital cannot be credited entirely to a single company or device. Nonetheless, it was the 1981 Sony Mavica that is recognised as being the first commercial digital camera.

At the time of its release, film-based systems dominated the stills camera market completely, but by borrowing some of the ideas that were starting to gain traction in the video camera market, Sony was able to deliver a completely new type of camera. The Mavica prototype was announced to the world at a packed press conference in Tokyo in August 1981, and the business of making cameras was changed for ever.

The name Mavica is a contraction of Magnetic Video Camera, and the camera's technology traced its ancestry more to the world of video cameras than to the film-based systems that dominated the stills camera market so completely at the time.

By borrowing some of the ideas that were starting to gain traction in the video camera market, Sony was able to deliver a completely new type of camera.

The Mavica built on the magnetic film-based storage that had already been in use in video cameras for years, and featured a capturing chip called a charge-coupled device, or CCD – a digital 'black box' that still sits at the core of most of today's digital cameras.

It was officially described as a still video camera, as technically images weren't stored in digital form, but rather on two-inch floppy disks. But, given that Sony had played a leading role in CCD development, and that the Mavica clearly set down a marker that other companies were quick to follow, there's no question that it deserves to be called the 'pioneer of the digital era'.

Commercial impact

Digital cameras are universally known and loved around the world today. Global Industry Analysts estimated in early 2011 that by 2015 the global market for digital cameras will exceed 138 million unit sales. But, as with so many emerging technologies, the early days weren't much to write home about from a commercial point of view.

The original Mavica, for instance, never actually went into mass production, although it did garner plenty of interest for the potential it offered. The analogue NTSC images were difficult to work with, only 50 could be stored on each disk and the device had to be connected to a TV to be able to view recorded images at all.

It wasn't until the early 1990s that the digital camera market began to gather pace. The start of the 1990s saw the arrival of the Model 1 from unheralded Dycam, which officially lays claim to being the first truly digital camera to arrive on the market.

The original Mavica ... never actually went into mass production, although it did garner plenty of interest for the potential it offered.

It may only have been capable of taking black and white shots, and had just one megabyte of internal storage, but at £499 it finally saw non-film cameras start to make a bid for the mainstream. Following hot on the Dycam's heels was the Kodak DCS100 – another all-digital system which showed that established camera makers were starting to take digital camera technology seriously.

The DCS100 was built around a standard Nikon F3, with the processing and storage happening in a separate box that had to be shoulder-mounted. It

The Kodak DCS100.

was far from elegant, and unsurprisingly was never a mass-market commercial success, yet for the first time it gave serious photographers the option to 'go digital' without any compromise of image quality.

The problem, though, was that all these early devices were niche products aimed at professional photographers, and there was little to suggest that the real future of digital cameras lay with the mass market. A company that played a key role in changing that was Apple, with its QuickTake line of digital cameras; a product range that suggested digital cameras could be neat, attractive and simple to use.

But the QuickTake didn't last long – image quality was judged to be dreadful, it could store a mere eight images and only worked with a Macintosh computer. Steve Jobs pulled the plug on the project three years after it was launched, and Apple hasn't made digital cameras again. What the QuickTake did achieve was the awakening of more established camera brands to the idea that digital could be marketed to the mainstream, and that really got the ball rolling.

In the late 1990s Sony's Cyber-shot brand arrived on the scene, aimed at the fast-growing consumer market. Despite being locked into several proprietary technologies, such as the MemoryStick storage card system, Cyber-shot was a bestseller for Sony around the world.

Sony soon entered another emerging market in digital single-lens reflex cameras (DSLR), a technology that combines a traditional viewfinder with digital technology, and bought out industry heavyweight Minolta in 2005. Sony is now the world's third-biggest digital SLR brand, behind only Nikon and Canon, while thanks to Cyber-shot it now trails only Canon overall – and is closing in all the time. The aforementioned report by Global Industry Analysts predicted that DSLR will drive the growth of the market and of unit sales globally by 2015.

What happened next?

Today, film cameras are little more than a relic of a bygone age and much of the camera market has moved to digital. Not only that, but the electronic building blocks of mass-market cameras have become just another component, and as a result have found their way into everything from baby monitors to computers, and of course the huge mobile phone market. Although they are traditionally poorer quality than dedicated cameras, the increasing sophistication of lenses on smartphones means that they are increasingly rivalling their dedicated counterparts.

The rise of the consumer digital camera has had a profound impact on the photo processing industry. The once-ubiquitous photo processing services on the high street have been all but wiped out of existence by the prevalence of digital, which allows consumers to view, edit and print their photographs from home, cutting an external service out of the equation.

Another device that has seen its demise in the wake of digital is the once hugely popular instant film (popularly known as Polaroid) camera. The ability to print your photos instantly was no longer a novelty, and Polaroid discontinued production of its devices in 2008, although commemorative versions have been released since then.

The rise of the consumer digital camera has had a profound impact on the photo processing industry.

39

Remote keyless entry (RKE) systems

When: 1983

Where: USA

Why: Remote keyless entry systems allowed car owners to secure their vehicle at the touch of a button

How: Inspired by the garage door openers of the 1950s, unique code technology was developed to increase security

Who: General Motors and Renault, although improved by Nanoteq in South Africa

Fact: The KeeLoq encryption system selects a code from 4,294,976,295 possible combinations

Remote keyless entry (RKE) systems are so ubiquitous nowadays that marketing a car without the feature would be seen as archaic. A business idea that combines ease of use with low cost – and most importantly of all, security – the remote keyless system has made entry and exit a breeze.

The background

Modern iterations of remote entry technology can be traced back to the 1950s, when two inventors completely unknown to each other both developed the first remote garage door openers in Washington, DC and Illinois. These first systems – despite being lauded as space-age – were in fact very simple, consisting of nothing more than a small, low-powered radio transmitter and a corresponding receiver, which opened the garage door.

A fundamental security issue was identified – the radio signals used by the devices were all the same, meaning that anyone who owned a transmitter could open anyone else's garage door.

This was perfectly adequate when such a product was an expensive luxury, but as more people bought the garage door openers a fundamental security issue was identified – the radio signals used by the devices were all the same, meaning that anyone who owned a transmitter could open anyone else's garage door.

It was well into the 1970s before companies began introducing basic security technologies into their garage door openers. These primitive measures were based on several hidden switches that matched the configuration of the receiver, allowing for about 256 possible combinations – enough to stop your neighbours from raiding your garage perhaps, but still not enough to be seen as a viable replacement for a key.

Over a decade later, the first keyless entry systems began to appear on new vehicle models, with General Motors and Renault offering a version of keyless entry on some of their models as early as 1983. Although the security had been stepped up from switches on these first devices (the internal codes were unique to each car and much longer), they were very expensive and their range was poor. In addition, criminals soon began to exploit a gaping security flaw – inexpensive, covert devices known as 'grabbers' could intercept the radio signal being transmitted from the fob and replicate it, essentially meaning that they had an instant clone of your car key and could easily gain entry to your car.

The Renault 25 introduced remote control central locking, operated by an infrared transponder.

Clearly, the technology was not mature, and many still preferred the security and inconvenience of the traditional key to the risks and convenience of the remote version.

Commercial impact

The breakthrough that finally made keyless entry systems a viable alternative to the key came in the late 1980s – the introduction of 'code hopping' security technology. This technology was pioneered by two scientists working at a South African security company, Nanoteq, where they came up with the KeeLoq encryption technology that is still used by many vehicle manufacturers and security providers today.

Instead of using just one code, the KeeLoq transmitter and receiver both come with a chip containing an algorithm that generates a single-use code (typically 40 characters) out of over four trillion possible combinations. When the button on the fob is pressed, it sends this code to the receiver, which is expecting that particular combination and performs an action, such as opening a door. The code is then changed according to the formula, and the next time the fob is pressed a different code is sent out. This renders 'grabber' devices useless – if they replicate a code that was sent out by the transmitter, the receiver will just ignore it, because by then it is expecting a different code.

The technology was sold to technology firm Microchip Ltd in the 1990s for $10m. The exponentially improved security proved to be the final piece in the puzzle, becoming an industry standard within a few years, and being adopted by a range of vehicle manufacturers, such as Chrysler, General Motors and Jaguar.

What happened next?

Modern cars that are manufactured without RKE have become as much of a curiosity as the cars in the 1980s that did have the technology. The market penetration is complete and RKE is now installed as standard on the vast majority of new vehicles.

Manufacturers ... are beginning to offer entirely keyless car models, with the remote fob providing the only means of entry and ignition.

New remote entry fobs offer features such as remote ignition – meaning the car can be started and stopped without a key – and newer iterations of the remote entry system are also expected to be able to display information about the car, such as whether it is locked or unlocked. Manufacturers have such confidence in the mature technology that some are beginning to offer entirely keyless car models, with the remote fob providing the only means of entry and ignition – doing away with the traditional key altogether.

One technology that is also growing in popularity is passive keyless entry. This triggers a car lock to open when the transmitter is within a certain radius of the vehicle and to automatically lock when it leaves that radius, adding a further layer of convenience and security. The idea is highly popular, with a third of cars in Europe expected to ship with the feature by 2012.

Furthermore, the traditional form of the key fob is also facing a new challenger in the form of the smartphone. There are now applications available for these devices which use the phone's chip to replicate the authentication system used by keyless entry systems. This convergence may signal another change in the way we view locking systems.

Whatever the future holds for keyless entry, it will always take the credit for having completely revolutionised the way we enter our vehicles. The fact that the technology is increasingly being expanded into other areas, such as home security, is just further evidence that the revolution doesn't stop there.

40

Dyson (Dual Cyclone vacuum cleaner)

When: 1986

Where: Wiltshire, England

Why: James Dyson re-wrote the rulebook for the simple vacuum cleaner and became an inspiration for manufacturers, inventors, entrepreneurs and designers

How: The poor performance of Dyson's household vacuum cleaner led him to discover that a bagless model would clog up less easily. Nearly 10 years later, his Dual Cyclone design was selling successfully in Japan

Who: James Dyson

Fact: A Dual Cyclone vacuum cleaner accelerates dust particles to speeds of 924mph around the cone

Hoover's unique selling point was the alleviation of manual chores, an aspiration encapsulated in the slogan 'It Beats as it Sweeps as it Cleans'. Dyson's 'Say Goodbye to the Bag', then, captures how substantially the technology that had dominated the vacuum cleaner market since the turn of the 20th century was disrupted by the Dual Cyclone cleaner. The invention emerged when James Dyson realised that industrial cyclone technology could be miniaturised for the home. By dispensing with the porous bag (that quickly clogged, resulting in a loss of suction), the Dual Cyclone vacuum cleaner offered greater efficiency and a contemporary design.

The background

The roots of James Dyson's invention, patenting and subsequent commercialisation of a bagless vacuum cleaner lie in his successes and mistakes in an earlier venture. In 1971, when working on the 300-year-old farmhouse he had bought but couldn't afford to renovate, he became fed up with the shortcomings of the builder's wheelbarrow he was using. In all the thousands of years since its inception, he thought, nobody had ever stopped and said: 'I could design this thing better.'

Three years later he was in production with the Ballbarrow, a much more stable and user-friendly design that was an immediate hit with the gardening public. It was never taken up by the building trade however, where resentment was 'almost masonic', teaching him the important lesson that the entrenched professional will always resist innovation much longer than the private consumer.

Kirk-Dyson, the company making Ballbarrow, did well in the domestic market in the mid-1970s, but was overly indebted and got into difficulties when it tried to enter the American market. In 1979 Dyson was sacked from his own company, following a dispute over rights. Dyson wanted to hold on to the intellectual property he had created – something he advises designers to do by all means. His partners, including his sister Shanie and her husband, wanted to sell the company, and promptly did so. As a result, James Dyson didn't speak to his sister and brother-in-law for 10 years, and learnt the lesson, he says, that one should not go into business with relatives.

Dyson quickly realised that, although vacuum cleaner technology had been trusted for generations, it was fundamentally flawed.

In 2005, Dyson resuscitated the ball-versus-wheel concept, applying it to his vacuum cleaners and making them more manoeuvrable. But the invention for which he is best known, the Dual Cyclone vacuum cleaner sprang from the industrial processes at Ballbarrow, rather than from the technology. Its inspiration, once again, came from a domestic chore at another farmhouse, to which he had moved his family.

Frustrated by the poor performance of their old Hoover upright, the Dyson family bought a powerful cylinder cleaner but found that it too lost suction after a short period of use. Dyson started experimenting with emptied used bags and new bags, noting a rapid decline in suction after a short period of use: 'It had to be that the pores of the bag – which were meant to let out only air – were in fact clogging with dust and cutting off the suck,' he decided.

Dyson quickly realised that, although vacuum cleaner technology had been trusted for generations, it was fundamentally flawed. From that moment, he declared war on the entire industry that had built up around a failing product. However you spin it, he argued, the bag would always be the weak link.

If the nuisance had been limited to the domestic arena, James Dyson might have gone on living with it. However, his engineering breakthrough was stimulated because, at about the same time, he had to deal with a similar problem at work. A powder-coating plant had just been installed at the Ballbarrow factory. Excess powder was collected in the booth by something that in essence was a huge vacuum cleaner with screens instead of a bag. The process had to be stopped every hour to clear the screen, and this down-time was a costly problem.

On investigation, Dyson found that big industrial powder coating companies used a cyclone, a tall cone that spun the fine powder out of the air by centrifugal force. He obtained a quote to have one built at his factory, but the cost was beyond the means of the business. Instead, he set about making one himself and used as a model the nearest available working unit, at a nearby timber mill, where it was used to collect sawdust.

Armed with sketches he had made at the sawmill, he welded up a 30-foot cyclone from steel sheet and installed it above the powder-coating plant, after discarding the screen with its cloth filter. The next day's production went ahead without any stoppage, as any powder that escaped was drawn into ducting that forced it round the walls of the inverted cone: it then spiralled down to be collected in a bag at the bottom and reused.

James with the now-famous DC01.

This success stimulated Dyson's thoughts to return to the problem of his vacuum cleaner at home. This cyclone was like throwing away the vacuum cleaner bag and never having to replace it. There was no reason, he thought, why the cyclone should not work in miniature. The same night, in October 1978, he went home and ran the old Hoover Junior just to prove that the collection was the problem, not the suction. Then, he made a one-foot high version of the cyclone out of cardboard, attached it to the Hoover and set about vacuuming his house.

Dyson had learnt that in a cyclone (a cone with its point facing downwards) the air is forced in laterally at the top, forcing it to spiral down. A particle that is revolved inside a curved wall increases its speed three times, and if the diameter of the curve is reduced, as in a cone-shaped chamber, it will continue to accelerate. A cyclone in a vacuum cleaner accelerates the dust particles from around 20mph on entry to 924mph at the bottom, where it is travelling at 324,000rpm. The mass of the particles at this speed is increased and they are thrust to the sides of the cone. Thus, the air at the centre is free of all matter and can be exhausted through a chimney while the dust falls into a bowl. That is the principle that all Dyson cleaners use. James Dyson didn't invent it, but he engineered its application to the small-scale vacuum cleaner.

Commercial impact

In 1979 Dyson set up the Air Power Vacuum Cleaner Company in partnership with Jeremy Fry, and spent the next three years making prototypes of varying designs and using different materials. His approach has always been to make one change at a time, testing it to see if it improves the performance, so it took four years to reach a working prototype. One intractable problem, the different performance in the cyclone of different sized and shaped particles, required a number of modifications before it could be solved. The solution was to combine two cyclones, an inner, high-speed one to catch the finer dust and a slower, outer chamber to separate larger objects such as hairs and bits of paper. This is what's meant by a Dual Cyclone.

Always Innovating – The DCI 6, a handheld vacuum cleaner that doesn't lose suction.

In 1982 Fry and Dyson decided that they had an invention and decided to sell the licence to manufacture it, changing the name of their company to Prototypes Ltd. Failure in the UK and US markets – due largely to resistance to change on the part of traditional manufacturers – was followed in 1986 by success in Japan, where the G-Force cleaner was launched by Apex Inc. Despite its price of £1,200, it was successful as a niche product and the sale of the Japanese rights enabled Dyson to set up his own manufacturing company at Malmesbury in Wiltshire.

The robustness of its casing and the fact that you could see the dust collecting in its transparent chamber were among the many innovations that appealed to purchasers.

The factory and research centre opened in 1993 and the DC01 cleaner rapidly became a design icon. Its impact on the market was as much due to the model's bright colours as to its technology. The robustness of its casing and the fact that you could see the dust collecting in its transparent chamber were among the many innovations that appealed to purchasers. By September 1994, once the reluctance of the major retailers had been overcome, the Dyson had overtaken all other vacuum cleaner brands in the UK. In 2005 it became the market leader in the USA.

In December 2006, James Dyson was knighted for his innovative contribution to British industry. Continuing sales saw his personal fortune swell to an estimated £1.45bn in May 2011.

In October 2010, the Dyson Group reported a doubling of it profits to £190m and a turnover of £770m, attributed largely to international sales. It is targeting South America, India and China in the medium term, while building on sales in the UK and USA, its two main markets.

What happened next?

In the decades since the first Dyson vacuum cleaner was launched, the Dual Cyclone principle has been refined by the addition of smaller, high-speed cyclones to cope with greater volumes of air, known as Root Cyclone and used in models from DC07 onward. In 2005 the Ballbarrow concept was united to a vacuum cleaner in the DC15, which uses a ball instead of wheels to enhance manoeuvrability. Many variants have been added to the product mix, including hand-held vacuum cleaners (DC16) and animal hair cleaners.

Furthermore, the Dyson brand has expanded its range to encompass an innovative high-speed, energy-saving hand drier and a bladeless fan, the Air Multiplier, further cementing the brand's position in the global retail market.

Although the Dyson Dual Cyclone range is no longer the only bagless vacuum cleaner on the market, it has continued to lead the UK market and has become a major brand in Japan and Australia.

41

Tablet computers

When: 1989

Where: USA

Why: Tablet computing is starting to transform the way businesses operate

How: A range of new technologies combined

Who: Jeff Hawkins of Palm Computing, most recently Apple

Fact: Over a billion apps are downloaded for the iPad each month

Tablets are one of the surprise hits of the 21st-century consumer technology market. For many years they were seen as a halfway house and something of a compromise – not powerful enough to be a home device and too large to be truly portable. But since Apple launched its iPad, sales of tablets have taken off.

Apple, which has arguably the world's most enthusiastic and devoted customers, has captured the zeitgeist with the iPad. Since its launch, sales of tablets have gone through the roof, and while this iconic tablet has spawned many imitators, it remains the undisputed king of the tablets.

The background

It may be surprising to discover that the earliest electronic tablets can be found as far as back the late 20th century. Early versions could detect handwriting motions with an electric stylus. Many more attempts to create commercial devices were made in the latter part of the 20th century, though these made little impact. Small electronic tablet and pen devices were available throughout the 1980s and were used as organisers and notepads. Several of these, including Psion's Organiser and the Palm Pilot, achieved reasonable sales, but none established itself as a significant on-going product.

The Gridpad, created by Jeff Hawkins, founder of Palm Computing, brought handheld computing a big step forward in 1989. It ran a DOS operating system and a pen was used to input commands. The device sold for $2,370 and was bought by those involved in data collection – a work device, not a leisure one.

Throughout the 1990s and into early 21st century, computer companies were experimenting with touchscreen technology and companies such as Microsoft, IBM and GO Corporation all unveiled tablet devices. But neither business nor consumers bought these in any meaningful numbers; these devices failed to demonstrate a clear role for themselves. Most advances in portable devices served only to benefit the mobile phone market. In 2007, however, Apple launched the iPod Touch to mass critical acclaim and, by doing so, laid the foundations for the iPad in 2010.

The Gridpad, created by Jeff Hawkins, in 1989.

The Gridpad, created by Jeff Hawkins, founder of Palm Computing, brought handheld computing a big step forward in 1989.

Commercial impact

Despite the early tablet computers' failure to excite customers, when Apple founder Steve Jobs announced the iPad to the world, consumers went into a frenzy. Over three million units were sold in the first 80 days of its launch in April 2010. Stores sold out and customers couldn't buy them fast enough. By the end of the year, nearly 15 million had been sold, with demand for more still very strong throughout the world.

Apple has fostered a love and affection for its products that must cause great envy among other electronics manufacturers. It has also worked hard to position itself as the cool option in a world where technology had been thought very uncool. Apple is the brand loved by artists, musicians, some technologists and those at the cutting edge – or so the story goes. It was also aided by the fact that Steve Jobs was a gifted and charismatic speaker, who had constructed a compelling persona which mimicked that of the friendly guy at the internet café, rather than the CEO of a global conglomerate.

Following the launch of the iPad, other manufacturers launched their own tablets to try to get a piece of the action. Samsung, Toshiba, BlackBerry and others all attempted to emulate the success of the iPad, but with nowhere near the same level of success.

When the iPad launched, almost everyone loved it – but weren't quite sure what they would use it for. In the 15 months since, uses have sprouted up – people use it for giving presentations in meetings, for taking notes at meetings, and even for raising and sending invoices to customers. And of course for email, web browsing and entertainment, too. The iPad was so much fun to use that senior business-people the world over could tell instantly that this was a game-changing product.

Apple has fostered a love and affection for its products that must cause great envy among other electronics manufacturers. It has also worked hard to position itself as the cool option in a world where technology had been thought very uncool.

Today many forecasters believe that the majority of business computing will shift towards devices such as tablets within five years, driven by their ease of use and portability and enabled by seamless wireless access to storage, printing and links to the internet.

The other area of significant business impact is applications. Applications, or 'apps' as they are commonly known, came to the attention of many through social networks. Some networks, such as Facebook, opened up their systems to allow programmers to design apps that could work in their architecture. The industry has boomed as game makers, marketers, blue chips, hobbyists and others all tried to create the next big app. Apple has embraced this and says its users are now downloading around a billion apps per month. It alone is responsible for an app market worth more than $2.5bn – and the end is not in sight. Google's Android and the other operating systems all have their own equivalent 'app economies'. This is a boon for small businesses, some of which are becoming substantial in their own right.

What happened next?

The technological and commercial battles in this space are still unfolding and no one can be fully sure where they will lead. Currently, Apple is reigning supreme and the iPad is the tablet of choice for most. However, Android devices are considerably cheaper, and as this market expands rapidly it seems inevitable that other players will eat into Apple's market share.

It is very early days indeed for tablets, but they seem sure to be a greater part of the business world in the near future.

1990s

42

International overnight courier services

When: 1991

Where: University of Ghent, Belgium

Why: Urgent parcel delivery has transformed many industries that rely on the transportation of physical goods

How: Demand for ever faster courier services inspired a superior 'ready to move' status

Who: Mark Kent, professor of logistics at University of Ghent, spearheaded by FedEx Corporation

Fact: Over 15.6 million packages are delivered globally each day by UPS alone

A t one time or another we will all have needed to deliver a parcel, gift or document urgently. In pre-industrial times your options were a runner, a rider on horseback or a homing pigeon, with varied success rates. Foot messengers ran miles to deliver their tidings and many contemporary marathons are based on historic messenger routes.

Today, thanks to corporations such as UPS, FedEx and DHL, most parcels can be delivered between the USA, Europe and Asia overnight, transforming the efficiency of business services in these continents. Now an essential service for millions of offices, hospitals and fashion houses, overnight courier services have played a key role in fuelling the growth of international trade and business development.

The background

One could argue that overnight courier services have existed since ancient times – but only for short distances. In the UK, a same-day courier market stemmed from the London taxi companies in the late 1970s, soon expanding to embrace dedicated motorcycle despatch riders. Many regional courier companies sprang up around the country, gaining popularity with legal institutions that needed to transfer confidential signed documents between parties on strict deadlines. The success enjoyed by the early couriers persuaded many freelancers to join the market.

Over 12 million packages are delivered globally each day by UPS alone.

'*Courier Locator*' services ... allow customers to track their package along every step of its journey via GPS.

Of course, in many cases Royal Mail could provide the same service as a courier company. However, some time-sensitive organisations have become disillusioned with the national postal service, finding it too expensive and unreliable; furthermore, many are worried that strike action could delay the transfer of documents. Today, larger courier firms such as UPS, FedEx and DHL command a substantial share of the market because they are seen to offer a more secure, reliable service.

Their market position is no accident; these corporations have worked hard to develop benefits that can provide their customers with greater peace of mind. These include 'Courier Locator' services, which allow customers to track their package along every step of its journey via GPS. *Information Week* magazine described FedEx's 'Sense Aware' real-time tracking technology as one of the best ideas for solving business problems, calling the organisation one of the top technological innovators in the USA.

It is appropriate, then, that when it began operations in 1973, it was FedEx that spearheaded the first next-business-day courier service in the USA. Then, in 1991, for the first time its 'ExpressFreighter®' service guaranteed overnight delivery of parcels couriered between the USA, Europe and Asia.

However it was Mark Kent, professor of logistics at the University of Ghent, in Belgium, who first devised the idea that a 'ready to move' status could be adopted by hundreds of stand-by couriers – an idea reliant on independent contractors who are ready to burst into action according to customer demand, on any given day or time. UPS exemplifies this with its 'Next Flight Out' service, which, it claims, provided the first same-day service and now offers guaranteed next-day delivery by 8am.

Commercial impact

Overnight courier services are now integral to the success of the express delivery industry, one of the world's biggest and most resilient sectors. By 2003, the express delivery industry was contributing $64bn to global GDP and supported 2.6 million jobs around the world. By 2010, the express delivery divisions of DHL and FedEx recorded combined revenues of more than $37bn. Meanwhile, UPS and FedEx handle goods equivalent to 3.5% of global GDP between them.

Given that many companies around the globe now employ a just-in-time production strategy, keeping inventory to a minimum and relying on last-minute receipt of parts and materials, express delivery is now more important than ever; airports, harbours and rail terminals play a crucial role in keeping production lines going. In fact, it is estimated that Memphis International Airport, a major air terminus situated just miles from FedEx's head office, has a $20.7bn effect on the global economy.

As markets have grown, various industries have come to depend on the swift transfer of goods, including the medical sector, which often uses couriers to transfer organs for transplant. In an instance such as this, an efficient service can be life-saving.

The fashion sector, as a truly international industry, also relies heavily on the support of couriers to transfer in-demand stock for catwalks and magazine photo-shoots between London, New York, Paris and Milan. Many fashion publications employ several interns or junior members of staff in each department, just to pack up and send garments by private or in-house couriers. Indeed, couriers may also be used by fashion brands to transfer new stock from areas of affordable industry, such as India and Bangladesh, to the shop floors of Western nations.

> Given that many companies around the globe now employ a just-in-time production strategy ... express delivery is now more important than ever.

As well as operating their own aircraft, in real terms, one way that courier companies can provide their service is by tapping into the excess baggage allocation of commercial airline passengers. Some courier services offer heavily discounted airfares in exchange for the use of a traveller's baggage space. This allows them to transport documents quickly and easily through the customs office of a country. At the destination, the passenger simply gives the paperwork to a waiting customs agent, who processes it from there on.

The year 2011 marks 20 years since the first international overnight courier service was made available to the market. Following its innovative breakthrough, FedEx remains the leader in overnight shipping, although UPS is now the biggest parcel service in the world. Currently employing over 400,000 people to deliver 15.6 million packages daily to 8.5 million customers around the globe, many of them via next-day delivery, the UPS model is an excellent example of the success of overnight courier services. With revenue of $49.6bn (nearly £31bn) in 2010, it also shows that this is an industry that is profitable for both the courier businesses and the customers they support.

What happened next?

Although the market for overnight courier services remains consistent, wider acceptance of digitally processed important legal documents may well limit courier services' professional margins. UPS experienced a negative growth in employee numbers of 1.80% in 2010, which is relatively tame in the context of a recession; however, for a company with nearly half a million employees, that still equates to job losses for well over 7,000 people.

There are other factors at play, as FedEx reported higher fuel costs and declining demand cutting into its revenue by 20% during the last quarter of 2009, before experiencing a year-on-year turnover recovery of 18% in the first quarter of 2010. By 2011 the company was reporting total annual revenue of $39.3bn (nearly £24.1bn).

Consequently, the outlook for overnight courier services remains positive; however, the future may be sensitive. In the context of the global recession, overnight courier services will undoubtedly be affected as businesses experience changing trade and spending patterns.

43

The smartphone

When: 1992

Where: USA

Why: The smartphone has changed perceptions of mobile communication and provided business users with an office on the go

How: IBM unveiled a vision of a mobile phone-PDA hybrid, but Apple made it into a commercial phenomenon

Who: Various companies, primarily IBM, Apple and BlackBerry

Fact: Experts predict that 300 million smartphones will be sold worldwide in 2013

Many of the ideas detailed in these pages can trace their origins back decades, to innovations created in the Victorian era, if not earlier. Not so for the smartphone; in fact it's less than 20 years since this device's earliest ancestor was introduced to the public. The iPhone, surely the most famous member of the smartphone family, has only been around for four years, and most of its cousins are even younger.

However, during its short life the smartphone has already transformed the way people work, the way they shop, the way they communicate and the way they spend their downtime. It's disrupted traditional IT and communications industries, while creating satellite markets all of its own. And, by sweeping away the established the existing technological order, it's paved the way for countless equally innovative devices in the future.

The background

Essentially, the smartphone is the product of two very different technologies: the mobile phone and the personal digital assistant (or PDA). Both of these are, themselves, relatively recent. Although work on mobile telephones began as long ago as the 1950s, it wasn't until the 1980s that development really began to speed up; companies such as Motorola, with the DynaTAC, and Nokia, with the Cityman, produced handsets that fitted perfectly into contemporary corporate culture and created a burgeoning market for mobile handsets.

Motorola Mobility, Inc., Legacy Archives Collection, Schaumburg Ill.

Meanwhile, the PDA had evolved, thanks to a chain of inventions in the 1970s. In 1974 George Margolin invented a modular device that looked like a calculator but could be reworked into a keyboard. Four years later, a small company called Lexicon created a handheld electronic language translator with functions including interchangeable modules, databases and notepads. In the same year two young inventors, Judah Klausner and Bob Hotto, conceived a device which looked and felt like a calculator but could also be used to store essential details such as names, telephone numbers and basic memos.

By the early 1990s both PDAs and mobiles were becoming big business. For the PDA market, Hewlett-Packard launched the Jaguar in 1991, and Apple began work on the popular Newton, a device that translated handwritten notes into memos and gave users the choice of several simple games. Meanwhile the mobile phone was demonstrating ever greater potential; Group System for Mobile Communications (GSM) technology came on stream in 1990, heralding the dawn of second-generation (2G) phones, and the first text message was sent in 1992.

Against this background, Frank Canova, an IBM engineer, came up with the idea of adding organiser functions to a cellular phone, and created a team of 40 full-time staff to work on the project. The product, called Simon, was eventually designed to combine the functions of mobile phone, pager, PDA and fax machine in one device based on a microprocessor that held two forms of memory: ROM for the core computing functions, and PSRAM, a form of random-access memory, for the user's essential information.

Simon was way ahead of its time in several respects. The core functions included a series of rudimentary 'apps' such as a calendar, calculator, address book and global clock; the entry system was speeded up with predictive text capability; and the relatively old-fashioned keyboard was replaced with touchscreen technology, made possible by a 1½ inch liquid crystal display (LCD) screen. When the product made its debut as a prototype at 1992 industry convention COMDEX, it made an instant impression – one attendee was reported as saying it was the first time a company had placed a computer in a cellular phone, rather than placing a cellular phone in a computer.

Frank Canova, an IBM engineer, came up with the idea of adding organiser functions to a cellular phone, and created a team of 40 full-time staff to work on the project.

But as soon as the product was officially launched, in spring 1994, problems began. The shipping date, originally scheduled for April, was delayed as a result of bugs in Simon's cellular faxing capability. When it finally reached the shelves, consumers were underwhelmed; Simon was slightly too heavy to fit in the average jacket pocket, it broke easily, the screen was too small and the battery ran out after just an hour of constant use.

After months of low sales, Simon was withdrawn from the market – it is estimated that only 2,000 units were ever made. But, by proving that a mobile phone could be combined with a personal organiser, it provided a crucial stepping-stone towards the later emergence of smartphones. Over the next few years a wave of similar products reached the market, facilitated by the creation of the Wireless Application Protocol (WAP) standard for mobile phones in the late 1990s.

Prior to the introduction of WAP, a number of mobile companies had begun to explore ways to create handsets with wireless internet connectivity. As a result, myriad technologies had sprouted, with little common ground between them. In response, four of the leading market players – Nokia, Motorola, Ericsson and Phone.com – came together to create the WAP forum in 1997, in order to formulate a common standard for wireless connectivity.

The proceedings of the WAP forum led to the establishment of a clear set of internet protocols that could be supported by an array of mobile phone networks, and gave mobile developers clear specifications to follow. The forum's members also agreed on the creation of a new type of server, called a WAP gateway, which would take a web page and translate it into wireless content using mobile-friendly programming language. With this innovation, the WAP pioneers had created a bridge between mobile devices and the burgeoning internet, and allowed the smartphone to become a commercial reality.

Even before the WAP mobile standards were introduced in 1999, mobile manufacturers had begun to test the market with phones that offered basic internet connectivity. The year 1996 had seen the launch of the Nokia 9000 Communicator, which included an integrated keyboard capable of providing internet access. A year later, Ericsson had launched a concept phone known as Penelope, which enabled the user to send and receive emails; this was the first device to be openly branded as a 'smartphone'. However, following the release of WAP, the trickle of web-enabled mobile products became a flood.

In 2000 Ericsson, eager to capitalise on the widespread interest sparked by Penelope, launched the R380, an ultra-light and compact device that included a built-in modem and utilised a new mobile operating system known as Symbian, which was specifically designed for the smartphone. Two years later Palm, which had already built up a substantial following within the business community thanks to the Palm Pilot organiser, unveiled the Treo, which offered all the functionality of Palm's existing PDAs, with the added bonus of email access. Meanwhile Research In Motion (RIM), a young company headquartered in Canada, began experiencing considerable success with its family of BlackBerry smartphones.

RIM's first offering, the 850, released in 1999, looked and felt like a pager, and its scope was limited by today's standards – it even ran on AA batteries. But gradually RIM began to add new functions to the BlackBerry, such as international roaming and push emailing, which automatically forwarded a user's emails from their office server to their individual mobile device. The company also doubled the size of the screen and memory, and then it launched SureType, a keyboard standard that assigned two letters to each key, allowing the manufacturers to reduce the size so drastically that their device could fit in a user's pocket. In 2005 RIM's BlackBerry sales shot up by 46% to 3.2 million units – sales would not fall again until the first quarter of 2011.

Then, in 2007, the flurry of development around the smartphone reached a crescendo with the launch of the iPhone, perhaps the most recognisable smartphone on the market today. The iPhone was born out of Apple founder Steve Jobs's vision for a device that harnessed the potential of the touch screen and an array of additional features such as music and internet connectivity, while providing the on-the-go convenience of a mobile phone. In 2005 Jobs commissioned a top-secret development project, carried out in collaboration with AT&T Mobility, with no expense spared. From robot testing rooms set up to test the iPhone's antennas, to scale models of human heads used to measure radiation, Apple pulled out all the stops to make its smartphone a success.

The BlackBerry Bold 9900 is BlackBerry's latest attempt to keep up with the iPhone.

For the iPhone, Apple managed to rewrite its desktop operating system, OS X, for a mobile device little bigger than a man's hand, and the developers were able to harness the power of the Safari web browser. This meant that, rather than having to compress web pages into a smaller format, the iPhone was able to display each page in its full, original form. Using technology that had already been mapped out for tablet PCs, Apple transformed the touch screen experience, creating an icon-driven interface that allowed users to manipulate everything from music tracks and photo albums and Word documents, using only their fingers. And the whole process was super-fast, thanks to the advanced capability of the new ARM 11 chip.

Commercial impact

The release of the iPhone was swiftly followed by the Android, a device based on cross-industry collaboration and knowledge sharing. The Android was the first in a series of products developed by the Open Handset Alliance, a consortium of technology companies set up to create open standards for mobile devices; furthermore, the software was released on open source, so manufacturers of all shapes and sizes could develop their own Androids.

This open, cooperative arrangement has resulted in numerous Android variants flooding the market, taking the smartphone to even greater peaks of sales and popularity. Indeed recent reports predict that global smartphone sales will reach 420 million units this year, and more than a billion by 2016. According to a report released by research specialist Olswang in March 2011, the rate of smartphone adoption has increased sharply in recent months; over 20% of UK consumers now have a smartphone, and this percentage rises to 31% amongst 24- to 35-year-olds.

The smartphone has already brought seismic change to the technology and communications industries. According to Gartner, PC shipments experienced their first year-on-year decline for six quarters during the first three months of 2011 – and a significant factor in this fall has been the emergence of the smartphone, which offers all the core functions of a desktop computer with far greater freedom and flexibility. Meanwhile, many commentators believe the iPhone has prepared the ground for countless equally innovative mobile solutions in the future. In the past, the mobile carriers were the pivot of the industry, and manufacturers were under pressure to create products as cheap and cheerful as possible to appease them. But now, as *Wired* magazine put it in 2008, 'every manufacturer is racing to create a phone that consumers will love, instead of one that the carriers approve of'.

The smartphone has already brought seismic change to the technology and communications industries.

Furthermore, the current generation of smartphones has given birth to a whole new industry based around applications, or apps. The idea of an online marketplace for phone applications actually came from Marc Benioff, a former Apple intern who rose to become CEO of Salesforce.com. Benioff suggested the idea to Steve Jobs, who turned the idea into a commercial proposition that ran as follows: Apple would take downloadable applications built by third-party developers and publish them on a central marketplace known as the Appstore. If an app was downloaded, Apple would split the profits with the developer. As soon as it went live in summer 2008, the Appstore became a huge success; user downloads generated $21m in revenue for app developers in its first three days.

Since then BlackBerry, Nokia and the Android developers have launched their own appstores, and the app has become a global phenomenon. More than 10 billion apps have now been downloaded, covering everything from news and games to journey planners and price comparison. According to one research company, worldwide revenue from consumer apps will reach $38bn by 2015, and a whole new type of development company is springing up, from boutique specialists charging £50,000 for producing a single app, to DIY app-build kits that charge less than £100 per design.

By altering the fundamental rhythms of people's behaviour, the smartphone has also left its imprint on countless other sectors, nowhere more so than in the retail industry. In 2010 the value of global mobile transactions reached $162bn. Many believe that this inexorable growth will see the closure of dozens of bricks-and-mortar outlets, which lack the convenience of mobile shopping. Belatedly, however, retail firms are starting to wake up to the potential of mobile-commerce. In April 2011, a UK study found that 74% of retailers were in the process of developing a mobile strategy, predicated on the iPhone.

The business landscape has also been transformed. According to a survey of business executives released last year, 34% of people now use their smartphone more than their computer for work-related tasks, and the greater mobility and flexibility of the smartphone are expected to trigger a surge in mobile working as busy executives choose to work from home with all the functionality of an office desktop computer. As more and more employees begin working remotely, the issue of smartphone protection will become increasingly important, creating a huge new revenue stream for IT security companies; in fact, Global Industry Analysts predicts that the smartphone security software industry will reach a global value of $2.99bn by 2017.

What happened next?

The smartphone is such a young technology, it's difficult to predict where it will go next. At present, it seems likely that we will see increasing convergence between smartphones and tablets – the huge screens of the latest HTC and Samsung phones would certainly suggest a blurring of the lines. The market may even settle on a form of smartphone–tablet hybrid carrying all the core functions of a desk-based PC – creating the first-ever truly mobile office.

At the moment, however, the smartphone industry is embroiled in a full-blown form war, centred on operating systems: Apple's iOS is competing with Android, Symbian, and Nokia's in-house system for control of the market. Although Google and Apple's systems currently enjoy the lion's share of sales, and app selection, the conflict has yet to reach a decisive point. The future of the smartphone, in particular the technology it uses, could hinge on the outcome.

44

Search engines

When: 1993

Where: USA

Why: Search engines revolutionised our ability to navigate the ubiquitous and ever-expanding internet

How: The exponential growth of the internet meant a succession of innovators saw a need to help people find what they wanted

Who: A range of PhD graduates and computer scientists, not least David Filo, Jerry Yang, Larry Page and Sergey Brin

Fact: Google receives about a billion search requests every day

These days, we take search engines for granted. The issue of how to find the information that users really want is one that challenged computer programmers from the internet's inception. With more than 600 billion pages of information on the internet, and rising fast every day, the chances of people finding what they need without something like a search engine would be tiny. And the internet would not have grown into what it is now.

We would not be able to use the internet in the way we do, reaching for it from desktops, laptops and increasingly mobile devices multiple times every day. Our lives would be different. And of course Google, now one of the world's most valuable and best-known companies, would not exist.

The background

Back in the 1950s, the various groups involved in building the internet were faced with a problem: as the amount of information available grew, how would people find what they were looking for? Unless they knew the specific location of the file they wanted, it would be like searching for a needle in a haystack.

The solution came in the form of information retrieval (IR) systems, which searched documents for pre-specified keywords, known as 'metadata', building up a database of the results that users could then search through. One of the first systems to use IR was the appropriately monikered 'System for the Mechanical Analysis and Retrieval of Text' (SMART), developed at Cornell University during the 1960s.

It wasn't until late 1993 that someone had the bright idea of including a form, so that even people without extensive knowledge of programming languages could search for information.

By 1993, things were getting more advanced. Matthew Gray, a student at the Massachusetts Institute of Technology (MIT), came up with the idea of a 'robot' that would automatically trawl the web, looking for information. The World Wide Web Wanderer would save information from web pages, such as titles and a few keywords from the first few lines of text, putting them into a database, known as an 'index'.

While the World Wide Web Wanderer was by no means the first search engine, it wasn't until late 1993 that someone had the bright idea of including a form, so that even people without extensive knowledge of programming languages could search for information. JumpStation, which launched in December 1993, not only crawled and indexed the internet but allowed people to search it too, making it the first search engine of the type we recognise today.

Created by a young computer scientist called Jonathon Fletcher, JumpStation never had the capacity to create a lasting impression on the world – in fact, it overloaded Stirling University's servers on its first day. Another problem that limited the impact of JumpStation was its sheer novelty – no one really knew how to use it, let alone harness and develop it, beyond the confines of Stirling's computer science department. Although JumpStation was nominated for a 'Best of the Web' award in 1994, Fletcher was unable to secure investment for his venture, and discontinued it in the same year.

However, helping people to find information online was by now becoming a popular pursuit. Many of the early search-based resources, such as Lycos, Yahoo! and InfoSeek, were portals – all-in-one sites that combined search functionality with an email service, news, sport and weather updates, and myriad additional services. Of all the popular sites around at this time, AltaVista was the only genuine search engine; although it enjoyed considerable early success, reaching a peak of 80 million hits per day by 1997, it was eventually overshadowed by newer players on the market, and after various takeovers ended up under the Yahoo! umbrella.

Yahoo! had originally been given the decidedly less catchy title of 'David and Jerry's Guide to the World Wide Web', after founders David Filo and Jerry Yang. The pair started the company from a trailer while they were studying for their PhDs at Stanford University. The idea was to 'keep track of their personal interests on the internet', but the site swiftly gained popularity among Silicon Valley's close-knit internet community, and Filo and Yang incorporated it as a business in March 1995.

What set Yahoo! apart from its competitors was that it was more of a 'web directory' than a search engine or portal. Yahoo! relied on a team of human researchers to organise all the information into categories, which would then be used to help people find what they were looking for. That made finding information easier, but the results presented when someone performed a search weren't necessarily the most relevant.

While Yahoo! grew rapidly, the most game-changing search engine launch came in 1998, when two Stanford PhD students, Larry Page and Sergey Brin, unveiled a project they had been working on for a couple of years. Unlike its competitors' all-singing, all-dancing sites, the search page had a simple, minimalist design (Page and Brin later admitted that this was because they weren't very familiar with the process involved in building flashy websites). They named their site 'Google', a play on the word 'googol', a mathematical term for the number one followed by a hundred zeros, which they said reflected the seemingly infinite amount of information they were planning to organise.

Aesthetics aside, the main difference between Google and its competitors was how it 'ranked' the information. Other search tools based their rankings on how often a search term appeared on each site: the more times the search word appeared, the higher the site would feature in the eventual ranking list. This was playing into the hands of spam-fuelled sites that simply repeated the same words over and over again, so Brin and Page came up with a new formula – whereby each site would be ranked according to how many other sites linked to it. They developed a complex algorithm that took into account the number of 'backlinks', as well as how and where keywords were mentioned. That cut out the spam, because people were far less inclined to link to dubious websites, which meant that the pages nearest the top were more relevant.

Brin and Page called the new system 'PageRank' (in honour of the latter), and although it took a while for the engine to become established, experts were soon impressed. *PC Magazine* credited Google with 'an uncanny knack for returning extremely relevant results', just three months after its launch. The creation of multiple language sites helped Google to establish control of the search engine market at the start of the new millennium. In June 2000, Brin and Page signed a deal to provide default search listings for Yahoo!, cementing Google's position at the summit of the search engine league.

Google refused to rest on its laurels. The same year as the Yahoo! deal saw the introduction of Google Toolbar, a browser that allowed users to trawl the net without visiting the Google homepage first. Google also launched Chinese, Japanese and Korean language services and, perhaps most crucially, rolled

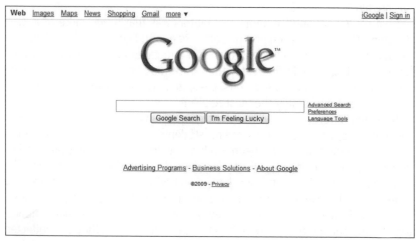

Google's classic homepage.

out a self-service advertising program known as AdWords. The venture was conceived around the time of the dotcom bust: Google's founders realised, in the run-up to the crisis in March 2000, that the site needed to find a way to make money. While Brin and Page had, until then, shunned the idea of advertising (thinking in horror about the eye-watering, flashing banner ads that dominated their competitors' homepages), they realised that they needed to come up with a way to sell ads that wouldn't ruin the minimalist aesthetic they had created for the site.

Google AdWords was a revolution. Users could activate their account with just a credit card, specify their own keywords, and tell Google how much they were prepared to pay for each click their ad received. If a user wanted feedback, they could download a tailored report or speak to a Google expert. Customers were wowed by this simple, personal approach to advertising – and, because their ads were being displayed against related search results, they knew that the people viewing the ads already had a genuine interest in what they were selling.

Commercial impact

Since establishing a control over the search engine market a decade ago, Google has become a global financial powerhouse; indeed the company's advertising revenues totalled $28bn in 2010. Furthermore, its ranking system has reshaped the advertising landscape.

In addition to the pay-per-click advertising sector, Google's PageRank innovation has given birth to a whole new industry: search engine optimisation (SEO). Brin and Page's PageRank algorithm remains a closely-guarded secret, which means that website managers don't really know what puts one page higher up the search results than another. But that doesn't stop them from trying to guess. These days, the quest to get websites to the top of Google's search results is big business; in fact, the SEO industry is now thought to be worth around $16bn.

Companies now spend millions on their websites, in the hope of getting a high Google ranking. Those that rank highly can realise huge benefits from the increased traffic and sales that their position brings. High-profile businesses, such as cheapflights.co.uk, have built their success on a high search engine placing, and whole industries are shaped by Google's ranking order. For example, it is estimated that 85% of internet users rely on search engines to access hotel information on the web, so well-designed, search engine-optimised websites are essential for hoteliers of all sizes.

The search engine revolution is likely to continue apace over the coming years. A report from Zenith Optimedia, released in July 2011, predicted that website advertising will rise 31.5% between 2011 and 2013, and much of this growth will focus on SEO and pay-per-click advertising as increasing numbers of companies come to see search engine marketing as a cost-effective alternative to traditional channels.

Companies now spend millions on their websites, in the hope of getting a high Google ranking.

What happened next?

As the rise of Google has continued, other search engines struggle to keep up. Yahoo! still has the second-largest share of search engine traffic, but in a recent interview with the New York Times, even CEO Carol Bartz admitted that the company has stopped trying to keep up. 'We were never a search company. It was more, "I'm on Yahoo!, I'll do a search",' she explained. Like many of its competitors, Yahoo! has branched into other technologies, transforming itself into a 'portal' that includes news, email and shopping channels.

Who knows what the search engine space will look like in 2021?

However, new challengers have emerged in recent years. Bing, the search engine rolled out by Microsoft in 2009, has grown steadily since its launch. In July 2011, it overtook Yahoo! as the UK's second most popular search engine. Microsoft has also inked a deal with Beijing-based search engine Baidu, which currently dominates the Chinese search engine sector; in fact it currently controls 74% of the Chinese market by ad revenue, with Google commanding just 23%. And in Russia, Yandex boasts a 64% market share of search engine traffic. As the most popular Russian website it attracted 38.3 million unique users in March 2011, and raised $1.3bn when it went public on NASDAQ in May 2011. The initial public offering was the largest for a dotcom in America since Google listed in 2004.

These figures show that Google, for all its dominance, still has plenty of competition, and must keep evolving to remain the market leader. But no matter what Google does, it's still far from certain that the company will be as dominant in 10 years' time as it is now. At the start of the millennium few of us used Google. Who knows what the search engine space will look like in 2021?

45

Pay-per-click advertising

When: 1996

Where: USA

Why: The pioneering method of payment transformed the emerging world of online advertising

How: By only charging companies when their ads were clicked through, pay-per-click payment models ensured effectiveness and value for money

Who: The founders of Open Text and, later, GoTo.com

Fact: In 2010, 4.9 trillion pay-per-click ads were displayed in the USA

They're easy to ignore, but those links that you see down the right-hand side and across the top of your internet searches haven't happened by accident. In fact, a huge amount of effort has gone into creating bite-sized ads specifically targeting what you're searching for.

Pay-per-click advertising, or PPC, works by charging the advertiser each time someone clicks on their link. The amount charged depends on the popularity of the search – the more common the keyword, the more it costs to get your ad displayed by the side of the search listings, and the more competitive it is too.

But there's no doubt that PPC is one of the most effective ways to get the word out about your company. For large firms, such as Amazon, that rely on it, it means millions of clicks a day. In fact, it's a revolutionary way of marketing – effectively guaranteeing to advertisers that every penny they spend will translate into a view of their website.

The background

When PPC advertising was first introduced in 1996, it caused an outcry among the users of the Open Text search engine, which pioneered it. People were scandalised at the idea that the ads were presented as text-only results: they could hardly believe anyone would try to confuse innocent web users by attempting to pass advertising off as real search results (despite the fact that they were clearly labelled).

But, as a model, PPC isn't completely new. In fact, the idea has been around for some time. Since the 1950s, savvy advertisers had been negotiating deals with publishers whereby they would pay the publisher a commission for every response they received to their ad in a newspaper or magazine. If they didn't get any responses, they didn't pay – exactly how the PPC model functions today.

As with many good ideas, though, the idea to show simple, text-based ads alongside search results was rejected outright after that outcry from Open Text users. In fact, for a while, advertisers reverted to brightly coloured (and, during the 1990s, increasingly irritating) banner and pop-up ads. These were sold on a cost-per-impression basis – that is, the advertisers paid up front for their banner to be displayed X number of times per month.

It wasn't until 1998 that the idea of paid search was resurrected by a little-known search engine called GoTo.com. At the time, GoTo sold all its listings to the highest bidder, based on the idea that the websites willing to spend most on appearing at the top would probably also be the most relevant to the user's search.

In hindsight, that model had various flaws – but GoTo's Eureka moment came when it decided that it would charge advertisers not for impressions, but for the number of people who clicked on their ads. The ad could be viewed by plenty of users, but unless it was relevant – unless someone clicked on the link

– the advertiser didn't pay a penny. Not only was it the ultimate indication that the ad was effective, but it opened the internet up to a whole new generation of small advertisers with restricted marketing budgets.

GoTo's problem, though, was that, as a search engine, it simply didn't have enough users to create the number of clicks that would translate into anything like the kind of money it needed in order to sustain itself.

But this was still the early days of the internet, and companies like GoTo were flexible enough to change their model entirely.

Thus, GoTo decided to do away with its search offering altogether, instead partnering with other search engines. Its idea was simple: the search engine would provide the results, while GoTo would provide the mechanism and software for advertisers to place 'sponsored' links at the top of the search listings, and they'd split the revenue between them. It worked. By 2001, GoTo was powering the sponsored search engines of MSN, AOL, Alta Vista and Yahoo!.

Google came up with a way to 'rate' ads and decide which ones should be at the top.

The one search engine it couldn't persuade to use its services, though, was Google. Despite its attempts at persuading Google to go into partnership, the search engine's founders, Sergey Brin and Larry Page, had clearly seen the potential of PPC and didn't want to share their slice of the pie.

Google's own offering, though, faltered at the first hurdle. Launched in 2000, it was still based on the classic banner ad payment model of 'cost per mille': advertisers paid per 1,000 impressions, rather than per click. After about six months of failing to persuade advertisers to part with their cash, Google realised that it was wrong, and decided to opt for GoTo's PPC model – with a few crucial changes.

The thing that set Google's new offering apart from its rivals was its bidding model – the amount advertisers bid to have their ads displayed at the top of the page. Google realised that users won't automatically click on ads if they're not completely relevant. In order for them to click on the link (and thus, for Google to get its revenue), they need to be displayed next to the most targeted results possible.

Google came up with a way to assign a 'quality score' to ads. Combining the keywords advertisers used in their copy, the quality (and popularity) of the page the advert linked to and the click-through rate, Google came up with a way to 'rate' ads and decide which ones should be at the top. An advert's ranking would depend on its relevance to a particular search, the relevance of the landing page it linked to, and the number of clicks it received.

Using an algorithm, Google then combined these three factors with its bidding system – so while advertisers couldn't simply buy their way to the top

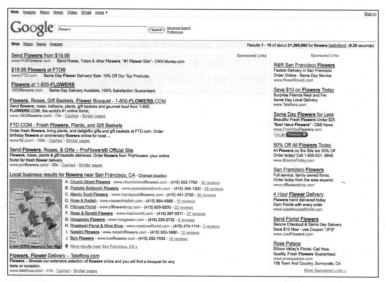

Google's PPC model is demonstrated here, with relevant links at the top and side panel of the search results.

of the page, they were still expected to pay a certain amount to come higher. The user found what they were looking for, the advertiser had their website viewed and Google got more in 'unrealised revenue' – everyone was a winner.

While its new AdWords offering caused a storm among online marketers in the early 2000s, it was the birth of Google AdSense in 2003 that really started to bring in the revenue. Google came up with a simple piece of code that allowed webmasters and bloggers – or affiliates, as they're known – to place ads on their own websites, thereby getting their own share of the revenue.

This innovation meant that businesses were no longer limited to simply advertising alongside search listings. Using the same system as its search engine listings, AdSense picked out keywords from the text of a website or blog and displayed highly targeted, relevant ads next to it. It caused a sensation – everyone from the smallest bloggers to some of the most popular websites on the internet began doing away with gaudy banner advertising when they realised they could generate cash using Google's discreet, more relevant ads instead. Bloggers, in particular, were highly impressed: here was a way to make money out of their hobby, with very little hassle.

Commercial impact

Having bought GoTo, later renamed Overture, in 2003, Yahoo! was reasonably confident that it knew how to maximise the financial potential of its new acquisition. And anyway, there was no way of telling how much Google was making out of its own offering, because it didn't publish its financial information.

The advantage for business owners with the PPC model was that, because it was based on clicks, rather than impressions, they could be confident their ad was working, every single time.

Until 2004, that is, when Google went public. As the rest of Silicon Valley continued to lick its wounds after the dotcom bubble's disastrous burst three years earlier, Google geared up to list on Nasdaq. There are reports of uproar in the Yahoo! boardroom when Google's first quarterly results came out: not only had AdWords gained considerable market share, but somehow, Google was managing to make three times the amount of revenue that Overture was making – per search.

There's no question that the marketplace was shaken by the revelation: not only did it prompt various search engines to switch from Overture, or Yahoo! Search Marketing as it was then known, to Google (by 2006 Overture had lost all its partners), but it also suggested to rival search engines that the only way they could maximise revenue from search advertising was by building their own systems. Suitably encouraged, Microsoft's MSN network began work on its own version of AdWords.

It was months, though, before Yahoo! announced that it would be getting rid of its old bidding system in favour of something decidedly more Google-esque. By the time Yahoo!'s version of the quality score system was launched, in 2006, the company was way behind Google in terms of innovation. And while Google had spent the time tweaking its algorithm, thereby increasing the amount of revenue it made, Yahoo! had merely been struggling to keep up.

It wasn't just big businesses that were affected by AdWords. Many small businesses' marketing options were changed for good, too. Now, instead of spending on ads in the classified sections of newspapers, or even big banner ad campaigns on websites, businesses could designate an exact amount to spend on search engine marketing – and when the limit they had set had run out, their ad wouldn't be shown any more. The advantage for business owners with the PPC model was that, because it was based on clicks, rather than impressions, they could be confident their ad was working, every single time.

What happened next?

Since AdWords' inception, Google has continued to hone its algorithm in order to keep ads as relevant as possible. This, in turn, has affected which advertisers make it to the top of searches. It has also spawned a new industry in search engine marketing. Companies charge a premium to help clients phrase their advert correctly, and will even recommend changes to a client's website to give it as good a chance as possible of reaching the top of the results.

In the UK, with consumers making almost five billion searches a month ... paid search marketing is now the single largest form of online advertising.

And while Google is now the clear market leader in search, capturing 65.1% of web queries in the USA in July 2011, competitors have been working hard to catch up. Microsoft has continued to enhance its search offering, culminating in the launch of Bing in June 2009, while a strategic alliance between Bing and Yahoo! appears to have enabled both players to strengthen their market positions. Comscore figures show that Bing commanded a respectable 14.4% of the US search market in July 2011, while Yahoo! maintained its second-place position, with 16.1%, up from 15.9% in June.

Meanwhile, search engines have become one of the most popular ways to spend advertising money. In 2010, the search engine marketing industry in the USA alone grew by 12%, as cash-strapped advertisers were attracted by the prospect of seeing exactly where their money was going. And this growth shows no signs of abating. A recent study conducted by Econsultancy and SEMPO (Search Engine Marketing Professional Organization) predicted that 2011 spending in the North American search engine marketing sector will grow from $16.6bn to $19.3bn in 2011, an increase of $2.7bn.

Meanwhile, in the UK, with consumers making almost five billion searches a month (according to ComScore, November 2010), paid search marketing is now the single largest form of online advertising, owning a 60% share of the market.

That's undoubtedly due, in part, to the tools that the likes of Google, Bing and Yahoo! make available to advertisers, allowing them to track everything from the number of clicks they receive every day to the most popular search terms, and even the percentage of clicks that translate into sales. In fact, advertisers are allocating greater proportions of their advertising spend to PPC every year, moving away from more traditional forms of advertising – in particular, local newspapers and printed magazines. PPC finally gives marketing executives the answer to their key question – which half of their advertising spend works? With return on investment easy to calculate, PPC seems set to remain an important part of the marketing mix for the foreseeable future.

46

Pay-by-swipe technology

When: 1996–97

Where: South Korea and Hong Kong

Why: Pay-by-swipe made travel ticket purchasing simpler, quicker and more convenient across the world's biggest transport networks

How: The Octopus card in Hong Kong was the first contactless smart card technology to be implemented across an entire transport network

Who: Mass Transit Railway (MTRC) Corporation and others

Fact: In Hong Kong, 95% of people aged between 16 and 65 have an Octopus card. As of August 2011, daily Octopus transactions are estimated at over HK$110m with over 11 million counts

When Hong Kong's major transport companies clubbed together to roll out pay-by-swipe technology across the country's entire transport network in 1997, it had a domino effect across the world. Similar payment systems for public transport now exist in Moscow, Washington, London and a host of other major transport hubs. Moreover, the idea of cashless micropayments is not restricted to buying travel tickets; phone manufacturers and banks are increasingly interested in micropayment technology for general retail purchases.

Hong Kong remains at the vanguard of the micropayment revolution, where the Octopus can be used for a vast array of small cash transactions – everything from buying groceries to paying library fines. Other developed economies are catching up fast, and a future without coinage is now considered a very real possibility.

The background

In the mid-1990s, Hong Kong badly needed change. As if adhering to the old British adage 'look after the pence', people in Hong Kong were hoarding coins embossed with the Queen's head, believing that they would appreciate in value over time. The lack of coins in circulation was making small cash transactions increasingly difficult. Particularly acute problems were experienced at Hong Kong's transport hubs, where millions of commuters needed to purchase small-denomination tickets for their journeys to and from work each day. The Hong Kong government needed a way of taking coins out of the equation.

Before the Octopus Card was launched, Hong Kong commuters pre-purchased their journeys on Common Stored Value Tickets, using a magnetic strip. In a move that would prove to be prescient, the Mass Transit Railway (MTR) Corporation had pushed for the Common Stored Value Ticket scheme to be extended to all other major transport networks in Hong Kong, creating ticketing uniformity, and a history of collaboration between the different networks long before the joint venture of the Octopus Card came to fruition in 1997. In 1994, MTR joined forces with Kowloon-Canton Railway Corporation, Kowloon Motor Bus Company, Citibus, and New World First Bus to form Creative Star Limited (later renamed Octopus Cards Limited).

The extremely successful Octopus card led the way for pay-by-swipe technology.

That MTR was and still is predominantly state-owned gave it considerable influence over other transport networks, which continues to this day. Indeed, MTR remains the largest stakeholder of Octopus.

Due to radio waves being used, swipe cards work within a range of a few centimetres of the reader, negating the need to actually swipe the card on the reader at all.

In 1994, MTR and four other major transport networks in Hong Kong founded a company called Creative Star Ltd. The company would later change its name to Octopus Cards Ltd and become a wholly-owned subsidiary of Octopus Holdings Ltd, allowing stakeholders to diversify product offerings and generate further profit. However, Creative Star Ltd was founded with the express purpose of modernising ticketing in Hong Kong and alleviating the change-shortage crisis. To that end, the company decided to implement a cashless ticketing system and tasked ERG Group, now operating under the name Videlli, to develop the technology for a swipe card payment system.

While Hong Kong's Octopus became the first large-scale roll-out of a cashless transport ticketing system, and the model that many Western countries looked to for inspiration, it was not, in fact, the first swipe card system. Seoul, capital of Hong Kong's close neighbour and economic rival South Korea, introduced its Upass swipe card system in 1996, a year before the launch of the Octopus.

The Upass scheme was introduced on a limited basis only, and has since evolved into the T-Money card, in widespread use across Seoul. Interestingly, Upass incorporated Philips' MIFARE classic 1k technology in its swipe card, whereas ERG Group chose to use Sony's FeliCa technology when developing the Octopus Card. MIFARE and Sony continue to be two of the leading technology providers for swipe card systems: MIFARE dominates market share in the USA and Europe; Sony has the edge in Asia.

All microchip technologies used in swipe cards, including those produced by MIFARE and Sony, work using radio-frequency identification (RFID). Radio frequencies are bounced between a reader and a microchip in the card, allowing for data to be read from and written onto the chip. Due to radio waves being used, swipe cards work within a range of a few centimetres of the reader, negating the need to actually swipe the card on the reader at all and explaining why some refer to the cards as 'contactless' smartcards.

RFID technology recognises when a commuter starts and finishes a journey, or makes a retail purchase, and deducts the appropriate amount from the card. Readers display the amount deducted and level of credit remaining on a card. Many swipe card systems currently in place require the cardholder to top up their card online or in person. Some swipe cards, such as the Octopus, allow users to set up an Automatic Add Value Service that tops the card up automatically once the credit drops below a certain level.

Commercial impact

In a sense, the Octopus was always destined to succeed in Hong Kong. Commuters were given three months to exchange their Common Stored Value Tickets for an Octopus Card, after which time the former became void. This contrasts markedly with Seoul's Upass card and later swipe card incarnations, such as the Oyster Card in the UK, where take-up was incremental and optional.

The advantages of swipe card technology are manifold. Most important for the cardholder is convenience.

It's no real surprise that three million people in Hong Kong signed up for the Octopus in the first three months. They had no choice. However, government influence alone does not explain the continued take-up of and affinity for the Octopus since its roll-out – 95% of Hong Kongers aged between 16 and 65 own an Octopus. But this doesn't account for the rapid implementation of swipe card technology in transport systems around the world, following the Hong Kong experiment.

Moscow Metro became the first transport network in Europe to start using swipe card technology, in 1998. Washington Metro Rail led the way in the USA a year later, with its SmarTrip Card. Boston and San Francisco both operate swipe card systems. London launched the Oyster Card in 2003, now used for more than 80% of tube journeys and 90% of all journeys on London buses.

The advantages of swipe card technology are manifold. Most important for the cardholder is convenience – one card with the potential to be used for a variety of different purposes. Commuters don't have to worry about buying individual tickets for journeys or having the correct change to be able to do so. Transport companies issuing swipe cards benefit from increased security and fewer fare dodgers. As most transport networks are heavily reliant on government subsidies, money that would otherwise have been invested in countering these problems can be given back to citizens in the form of lower taxes or fares.

Swipe card systems have also increased the speed at which commuters can move through barriers at rush hour – particularly important in busy metropolises such as Moscow, London and Hong Kong. The standard transaction time for transport readers in Hong Kong is 0.3 seconds. And the introduction of swipe cards has also cut down on congestion around ticket booths and the number of staff needed to sell tickets, further reducing overheads.

What happened next?

The infiltration of swipe card technology in cities such as Moscow, Washington and London is almost entirely restricted to the transport networks. As a business idea, therefore, swipe card technology is in an embryonic phase. To understand the full potential, one has to look eastwards. Nowhere has swipe card technology so completely permeated daily life as in Hong Kong.

Since the Octopus launch in 1997, retail outlets, wet markets, self-service businesses, leisure facilities, schools, parking meters and access points to buildings have all been adapted to incorporate its use. RFID microchip technology is now so small that Hong Kongers need not even carry a card. Microchips can be inserted into watches, key chains and ornaments, such products generating new revenue streams for Octopus Cards Ltd.

Taiwan and Japan have, in varying degrees, followed Hong Kong's lead. The Easy Card, introduced into Taiwan solely for transport purposes in 2002, can now be used to make small retail purchases. So too can the Suica Card in Japan. It seems almost inevitable that other countries using swipe card technology for transport purposes will eventually broaden its usage to include retail micropayments.

The UK has also begun making strides in this area. In September 2007 Barclaycard launched its OnePulse card, effectively a credit card with swipe functionality. OnePulse allows users to swipe for retail purchases up to the value of £15 without having to enter a PIN. Visa, Mastercard and American Express all now facilitate swipe functionality, but no other major UK card provider has yet taken advantage of it.

To understand the full potential [of the swipe card] one has to look East. Nowhere has swipe card technology so completely permeated daily life as in Hong Kong.

One possible explanation for the hesitancy in the UK (as well as the fact that the transport networks are deregulated) and elsewhere is fears over data security. In 2010 there was an outcry in Hong Kong when it emerged Octopus Cards Ltd had been selling on customer data to marketers. An extreme solution is Malaysia's compulsory MyKad identification swipe card, which contains a biometric fingerprint stored on the microchip and acts as an all-in-one ATM card, driver's licence and public key. However, this is unlikely to assuage the fears of those concerned about privacy. Most swipe cards in use are registered anonymously, few users choosing to give personal details.

The commercial advantages for suppliers and convenience for cardholders will most likely win out in the end. Sony's FeliCa microchip technology is already incorporated into its phones in Japan and there are suspicions that Apple intends to follow suit with its iPhone. Meanwhile, in the UK, Transport for London plans to introduce technology allowing commuters to pay for travel using contactless-payment credit and debit cards; it is that hoped the system will be up and running across the capital in time for the 2012 Olympics. Swipe cards may soon be a thing of the past, but swipe-and-pay technology has a long future ahead of it.

47

The hybrid car

When: 1997

Where: Japan

Why: The hybrid car has the potential to reduce the environmental impact of the automobile industry

How: Toyota developed Victor Wouk's hybrid power unit to launch the first commercial hybrid car

Who: Victor Wouk and Toyota

Fact: Hybrid cars like the Toyota Prius produce 90% fewer pollutants than comparable non-hybrid cars

Strictly speaking, the first hybrid vehicle may have been the Lohner-Porsche Elektromobil, which was shown at the Paris Exposition of 1900. Although it started purely as an electric vehicle, Ferdinand Porsche added an internal combustion engine to recharge the batteries, making it a true hybrid, by definition. The first half of the 20th century was dotted with attempts to combine technologies so as to achieve more efficient propulsion, but none got beyond the prototype stage and, more importantly, none went into production.

But if any one individual can be credited with the invention of a hybrid power unit, it should be Victor Wouk, the brother of the author of *The Caine Mutiny* and many other best sellers, Herman Wouk. For although his invention never went to market either, it was conceived with mass production in mind and developed for a commercial programme.

Toyota developed its technology independently and claimed – justly – to be the first to launch a commercial hybrid car, in 1997, by which time battery design and engine management software had reached the required level of dependability. Earlier attempts to produce a production hybrid, like the 1989 Audi Duo series, were based on separate electric and internal combustion drivetrains, so Toyota's claim to be the first to develop a production hybrid is unassailable.

The background

Victor Wouk was an electrical engineer and entrepreneur who, during the 1960s, devoted himself to combining the low-emissions benefits of an electric car with the power of a petrol engine to produce a hybrid vehicle. When he learned about the Federal Clean Car Incentive Program, run by the UC Environmental Protection Agency, he submitted a design and won a $30,000 grant to develop a prototype.

The car he selected as the vehicle for his power unit was a Buick Skylark, which he fitted with a 20-kilowatt direct-current electric motor and a Mazda RX-2 rotary engine. This vehicle was tested at the EPA's emissions-testing laboratories in Ann Arbor, where it returned a fuel efficiency reading twice as good as that of the unconverted car. Its emission rates were about a tenth of those of an average petrol powered car. Nevertheless, the project was dropped when

Wouk poses proudly with his 1972 hybrid Buick Skylark at the EPA test site.

it was realised that petrol was so cheap in the USA that a hybrid car had no market advantage, despite its environmental benefits.

Although the US government introduced further clean vehicle initiatives, notably the Partnership for a New Generation of Vehicles (PNGV) in 1993, it was left to Japanese manufacturers to develop and launch a production hybrid. In fact PNGV can be said to have handed the prize to Toyota, because it excluded foreign manufacturers.

The launch of the Prius in 1997 ... ahead of the Kyoto summit on global warming, can be said to mark the birth of the commercially viable hybrid car.

In the same year, and piqued by the US government's protectionist stance, Toyota pitched its research efforts into its secret G21 project. This accelerated product-development drive led to the Toyota Hybrid System (THS), which combined an internal combustion engine fuelled by petrol with an electric motor. The company's first hybrid 'concept' car was unveiled at the 1995 Tokyo Motor Show.

The launch of the Prius in 1997, two years ahead of schedule and ahead of the Kyoto summit on global warming, can be said to mark the birth of the commercially viable hybrid car, since Toyota sold 18,000 in Japan alone that year.

Commercial impact

At the time that the Prius emerged it was not clear whether hybrid technology would prevail over all-electric vehicles, and research funding was divided between these two mechanisms. However, the launch of the Prius tipped the scales and the first batch of electric vehicles, such as the Honda EV Plus, the Chevrolet S-10 electric pickup and Toyota's own RAV4 EV, were all eventually withdrawn.

In the year 2000, the Prius was introduced to the US and British automobile markets, and by 2002 more than 100,000 units had been sold. With a starting price of $22,000, this niche model proved a significant

TOYOTA — Toyota Prius - The world's most environmentally friendly family car   Issued 10/2001

The Toyota Prius, the first commercial hybrid car.

success for Toyota and soon a number of Hollywood names were seeking to boost their green credentials by getting behind the (hybrid) steering wheel, providing it with further positive publicity.

The launch of the Prius in fact had a stimulating effect on the entire hybrid automobile market. Research into electric and electric-hybrid vehicles continued and the development of batteries, in particular lithium-ion batteries, made hybrids more efficient. As well as this, the Prius levelled the playing field between hybrid and all-electric vehicles, so it is now an open question, just as in the 1970s, as to which of the mechanisms will prevail. Nonetheless, the Prius has a strong body of supporters, with more than 2.2 million vehicles sold globally by May 2011.

What happened next?

It was not long before Toyota's hybrid had a rival on its tail, in the form of Honda's Insight vehicle, in 1999. These Japanese manufacturers have held sway internationally ever since, despite the efforts of Ford (who produced the first hybrid SUV, the Escape Hybrid, in 2004), Audi and GM.

Worldwide sales of hybrid vehicles have been growing fast in recent years and Toyota remains the market leader – but more manufacturers are entering the race. At the 2010 Geneva Motor Show Volkswagen announced the launch of the 2012 Touareg Hybrid, as well as plans to introduce diesel-electric hybrid versions of its most popular models in the same year. However, the Peugeot 3008 Hybrid4, due to be launched in the second half of 2011, is expected to become the world's first production diesel-electric hybrid. Audi, Ford, Mercedes-Benz and BMW all have plans to launch hybrids before the end of 2011.

These projects appear to reinforce that hybrid cars occupy a secure market niche today, principally among people who want an environmentally friendly vehicle. It should be noted, though, that they still only account for just over 1% of global car sales, and fears remain that improvements in batteries, motors and, above all, the infrastructure for recharging electric cars may erode – and eventually wipe out – the hybrid electric market. Nonetheless, the legacy of hybrid cars, as a significant step forward in the progress towards environmentally friendly travel, will endure.

Hybrid cars occupy a secure market niche today, principally among people who want an environmentally friendly vehicle.

48

The MP3 player

When: 1998

Where: USA and South Korea

Why: MP3 players, particularly the iPod, have transformed the music industry, ushering in today's world of downloads and file sharing

How: The early MP3 players began the digital music revolution, but then Apple released the iPod, eclipsing all that came before it

Who: SaeHan Information Systems, Diamond Multimedia ... and, most significantly, Apple

Fact: The iPod sold over 100 million units in the six years following its launch

Walk onto any bus or train nowadays and you are certain to see at least a few pairs of the iPod's distinctive white earbuds adorning the heads of passengers. Yet just 10 years ago, the MP3 player was a device reserved for the serious technophiles of this world: lacking the essential features of platform support, competent design and ease of use.

Apple changed all that with a product that successfully utilised accompanying software as well as having a clean, attractive design to create the first MP3 player with mass-market appeal – and propelling Apple itself to unimaginable new heights. A business idea that saw the merit in making technology beautiful, the iPod has changed the way the world listens to music.

The background

The history of the iPod can be traced back to 1987, when German company Fraunhofer-Geschellschaft began development of project EUREKA. As discussed elsewhere in this book, the Walkman had created an obvious market for music on the move; aiming to build on this success, EUREKA strove to find a way of converting digital music files ripped from a CD into files that were much smaller, but without the huge compromise in quality that this would normally entail.

Headed by the scientist Dieter Selzer, who had been working for some years on improving vocal quality over a telephone line, the standard that later became MP3 was based upon a technique called 'perceptual coding'; certain pitches deemed inaudible to most listeners were 'cut off' the digital waveform, dramatically reducing the file size. Fraunhofer estimated that a file encoded in this way could be 12 times smaller than a song on a CD, without any reduction in quality.

The technology was submitted to the Moving Picture Experts Group (MPEG), which had been commissioned by the International Standards Office to find a new default standard for digital music files, and duly in 1991 MPEG-3, or MP3, was born, with the standard receiving a US patent in 1996. Two years later, Fraunhofer began enforcing its patents and all MP3 encoders and rippers were subsequently required to pay licence fees to the company.

Other digital music formats quickly appeared, building upon and improving the advances made by Fraunhofer. These included the Advanced Audio Codec (AAC), a higher-quality format officially recognised as an international standard in 1997.

Despite the standard being finalised in 1991, software programs utilising these new ergonomic formats were slow to appear. Fraunhofer released media player software in the early 1990s, but it was a dismal failure. The first popular digital music player software for PC was the free Winamp for Windows, released in 1998. Around this time, the first portable Digital Audio Players (DAP) also appeared on the market.

The Rio PMP300, retailed at #200, was notorious for falling apart, and ... users could store only around eight songs on it.

Early MP3 players included MPMan, developed by South Korean SaeHan Information Systems, and the Rio PMP300, produced by Californian firm Diamond Multimedia. These pioneering models attracted considerable interest; however, the early adopters of portable MP3 players had to deal with a number of glaring flaws.

The devices were either huge and unwieldy, or small and virtually useless; the interfaces were esoteric and difficult to use and, crucially, the accompanying software was either non-existent or highly inadequate. Not only did this put the MP3 player out of the reach of the layman, it also meant that users often resorted to embryonic file-sharing services such as Napster, fuelling the boom in piracy that was beginning at the time.

Each device had its own shortcomings; for example, the Rio PMP300, retailed at $200, was notorious for falling apart, and shipped with only 32 megabytes of internal memory, meaning that users could store only around eight songs on it.

Californian technology company Apple observed these flaws and began thinking about developing a portable media player that would improve on the usability and capacity of previous devices, to bring MP3 players to the wider market for the first time. It began working with hardware company PortalPlayer in mid-2001 to create the device, which was to have a hard disk to increase the number of songs that could be stored, as well as to work seamlessly with the iTunes media player software, released on Mac a few months earlier.

The project was shrouded in complete secrecy and Apple's then CEO and founder Steve Jobs was reputed to be obsessed with it, devoting 100% of his time to its development. Prototypes were encased in secure, shoebox-sized plastic casing, with the screen and buttons placed in random positions to ensure that the real design remained a mystery.

On 23 October 2001, Apple announced the impending release of the iPod. The first iPod had a five-gigabyte hard disk, enabling it to hold about 1,000 songs, a simple, stripped-down user interface with an innovative scroll wheel for navigation and the ability to sync with Apple's own line of Mac personal computers through iTunes.

The product initially met with some scepticism from the press, which pointed to the rather high retail cost (around $400), the fact that it worked only with Macs and the scroll wheel. The wheel was seen by some as an unnecessary gimmick. Nevertheless, the product began to grow in popularity, with users drawn to the attractive, clean design and the ease of use of the device, as well as the potential for carrying an entire library of music around with you. Just a few weeks after the European launch in November, the first unofficial tools for syncing the iPod from a PC began to spring up, a signal of the wide, latent demand for the device.

Taking note of this, Apple released the second-generation iPod in 2002 with added compatibility with Windows PCs, as well as a touch-sensitive scroll wheel instead of the manual kind found on the first-generation device. These later iterations also added more

The iPod Classic – a clean, attractive design.

capacity and greater battery life. In 2003 the iPod Mini was released, a smaller version of the vanilla iPod, designed to take on the high-end Flash-based music players with which the iPod was competing. However, the greatest leap forward for the iPod during this period was not a hardware feature but the addition of the iTunes Music Store in summer 2003. Integrated with the eponymous media player, the Music Store sold individual tracks for download at 99 cents and albums for $10, allowing users to access a world of instant music legally.

Later iterations of the iPod added video support and capacities of up to 160 gigabytes, and Apple began to focus on diversifying the product line past the basic model, with the company now offering Shuffle, Nano, Touch and Classic versions of the device.

Commercial impact

A host of renowned manufacturers have tried to knock Apple off its perch at the top of the MP3-player market, but most attempts have been fruitless. One of the biggest failures has been Microsoft's Zune, a line of digital media players that seemed to imitate the iPod in several key respects, notably its rectangular shape, large screen and scroll wheel. In July 2010, it was found that the iPod commanded 76% of America's MP3 market; Zune had just 1%.

While the iPod has established pre-eminence in the digital audio player market, iTunes has become the world's biggest music store, having passed the remarkable landmark of five billion track purchases in June 2008. In June 2010, iTunes occupied over 70% of the downloads market in the USA, and 28% of music sales overall.

The iPod, iTunes and their rivals in the MP3-player market have played an instrumental role in re-shaping the music industry, creating a new user relationship based on digital downloads rather than physical copies. In America, for example, digital album sales increased by almost 17% in 2010, while sales of physical albums declined. This trend has underpinned the growth of download sites such as Napster and eMusic; on the other hand, it has heralded a period of anxiety and drastic restructuring for traditional music stores.

Hundreds of independent record shops have had to close down, and even the industry's biggest names have proved vulnerable. In May 2011 British retail giant HMV announced a plan to close 60 stores over a 12-month period, in response to a sharp fall in sales; meanwhile Virgin Megastores has closed down in several countries, including the UK, America, Spain and Australia.

The MP3 player has also fostered a huge illegal industry, based on piracy. In 2008, the International Federation of the Phonographic Industry estimated that 95% of music downloads were illegitimate, and such rampant boot-legging played a significant role in damaging the music industry's overall revenues, which fell by 4.8% in 2010. Over recent months, many successful recording artists have been pressured into cutting the prices of their releases so as to compete with sites that share them for free.

The iPod, iTunes and their rivals in the MP3-player market have played an instrumental role in re-shaping the music industry.

What happened next?

Having conquered the portable media-player world and revolutionised the way in which many people download music, Apple has shifted its focus. It has taken on the portable device market in general, beginning with the launch of the iPhone in June 2007 (covered elsewhere in this book). The iPhone combines the features of an iPod with a phone and camera, and it has also enjoyed phenomenal success, with sales passing 100 million in June 2011.

However, more recently iPod sales have begun to plateau. In fact, a report in April 2011 revealed that sales were down 17% year-on-year. Industry analysts believe this slump is largely attributable to the rise of streaming sites such as

Spotify, which allow users to listen to a song straight away, without having to download or buy it. Many also believe that Apple, by pricing the iPhone at the same level as the iPod, has played a key role in damaging the latter's popularity; given all the extra features available on an iPhone, it's easy to see why someone would prefer it to an iPod, for the same cost.

Now that Spotify has moved into the American market, and smartphones are gaining new features all the time, the iPod could face a real battle to retain its market dominance. In fact, the future of the MP3-player market is far from certain. However, no matter what happens, few would dispute the critical role of the MP3 player in shaping the music industry as we know it today.

2000s

49

Google's 20% innovation time

When: 2001

Where: USA

Why: 20% innovation time has facilitated some of Google's most ground-breaking innovations

How: Google realised the smart ideas would come from the bottom, not the top

Who: Eric E. Schmidt, Larry Page and Sergey Brin

Fact: In the five years following Schmidt's appointment as CEO, Google grew from 5,000 to 9,400 employees and from $3.2bn turnover to $23.7bn

I n the early 1990s, despite all the advances made by management gurus such as Ken Blanchard, Spencer Johnson and Percy Barnevik, none of the world's major organisations had really cracked the problem of how to retain top talent. Some companies had begun to harness the concept of 'empowerment', whereby employees are encouraged to buy into goals. But for many employees, this concept wasn't enough; hundreds of companies were suffering because they simply couldn't hang on to top talent.

The Pareto principle, which had been coined by management guru Joseph M. Juran in the 1940s, suggested that 80% of a company's most productive ideas come from just 20% of its employees. Recognising this, other organisations, among them The Timken Company, allowed development groups to set up autonomous arm's-length entities that were financed, but not directed, by the company as a way of keeping products and intellectual property in-house. However, Google has taken the concept of employee autonomy onto a new plane – and the process began with the appointment of Eric Schmidt as CEO in 2001.

The background

Before he arrived at Google in 2001 to serve as a mentor for the youthful Larry Page and Sergey Brin, Eric Schmidt was not well known beyond Silicon Valley. But his unconventional ideas suited Brin and Page.

Schmidt argued that chaos was crucial to Google's corporate culture – indeed he once said that 'the essence of the company is a little bit of disorganisation, because it allows it to see what's next'. From this premise he devised Google's own 70-20-10 rule, at the centre of which lies the 20% innovation time principle – everyone at Google, from Page and Brin to the rawest Noogler (newly hired Google employee), should spend 70% of their time at work on core tasks, 20% on 'projects related to the core business' and 10% on new projects that aren't necessarily dedicated to day-to-day business processes.

When Eric Schmidt (left) joined Google as CEO, innovation within the company created by Sergey Brin (middle) and Larry Page (far right) flourished.

Shannon Deegan, director of people operations at Google, explains: 'You can take 20% out to work on anything you think is cool. If you think you have a great idea you can gather support from other people and maybe do some more work on it together.' There's a place on the Google intranet where these ideas can be posted to attract the critical attention and cooperation of colleagues with supplementary skills. However, there are some boundaries; Megan Smith, vice president for new business development and general manager of the philanthropic Google.org, has said: 'The 20% is not there so that people can entirely do their own thing; it's designed to make the company and the world a better place.'

Since Google works from the bottom up, great technical ideas are not devised in a separate department, with resources directed by management to develop them. The first task is to take the idea to fellow engineers and convince them that it has legs. This approach ensures against technical mistakes but also means that the burden falls upon the individual Googler to spread the idea.

Ideas that look as if they are going to have an impact on the way the company is run or perceived by potential employees are then developed within a 'grouplet'. These have neither budget nor decision-making authority, but bring together individuals committed to an idea and willing to work in order to convince the rest of the company to adopt it. Rather than resulting in a new product, many of these ideas have affected the way the company works from day to day. The shuttle buses that take staff to and from the company's headquarters in Mountain View, California are an example of this, as are 'Fixit Days' devoted to concentrated work on an engineering, customer relations management or documentation problem, and 'Testing'. This last one proposed that engineers should devise automated tests along with the products they devised – and disseminate them by posting them in the toilets.

'The 20% is not there so that people can entirely do their own thing; it's designed to make the company and the world a better place.'

Commercial impact

Many of Google's most effective customer-facing technologies have their origins in 20% time, including Gmail, Google News, Google Sky, Google Art Project, AdSense and Orkut. Since 99% of Google's revenues come from advertising, every application that increases the number of views on a Google platform has contributed to its market dominance, and 20% time has ensured that the 'cool' ideas that no boardroom would ever countenance get to the market fast.

Orkut, the social networking site developed in 20% time by Turkish engineer Orkut Büyükökten, grew to a user base of 120 million by 2008. The idea for Google Sky was floated by a group with an interest in astronomy who said: 'Wouldn't it be cool if we turned those Google Earth cameras up to the sky!' Working in 20% time they came up with a product that allows users to point their mobile phone to the sky and be told exactly which stars they are looking at.

Gmail, said now to be the world's third largest email platform, with over 193 million users every month, was developed in 20% time by Paul Buchheit and a group of associates, originally as an internal system, and was launched publicly in April 2004.

Given these successes, one might expect other companies to adopt the 20% innovation approach. However, in reality, this is impractical for most firms; businesses that lack Google's resources and market dominance can't afford to give their staff such freedom. A number of companies have taken similarly forward-thinking approaches to innovation – for example, Facebook empowers its engineers by giving them complete control over the projects they work on. However, the concept of 20% innovation has yet to enter common use around the business world.

Gmail, said now to be the world's third largest email platform, with over 193 million users every month, was developed in 20% time.

What happened next?

As Google became bigger it found it harder to retain its brightest employees against the blandishments of even 'cooler' companies such as Facebook. In November 2010 it raised compensation by at least 10% across the board as part of Eric Schmidt's 'war for talent'; it also started to consider establishing an in-house incubator, conceived as '20% time on steroids', to give people the chance to devote all of their time to innovation.

Larry Page took over as CEO of Google in 2011, with Eric Schmidt moving aside to the post of executive chairman. Schmidt has been using his own 20% time to investigate how technology can change the public sector and international relations.

50

The e-reader

When: 2007

Where: USA

Why: The advent of e-reader technology is transforming the publishing industry and driving the decline of printed book sales

How: Development of screen technology in e-readers led to e-ink (electronic ink), which was later combined with Wi-Fi to create a superior device

Who: E Ink Corporation and Amazon

Fact: The first-generation Kindle sold out within five and a half hours

E-readers have been available for more than five years, yet consumer interest was relatively lukewarm – until the Kindle was introduced into the market. While Amazon's e-reader range dominates the market, there are numerous others providing healthy competition, plus e-book offerings on tablet computers.

The incredible rise of e-reader technology shows just how powerful it is as a concept. It already threatens to replace the oldest and most durable form of media in history.

The background

Perhaps the most remarkable thing about e-readers is how long they took to become mass-market alternatives to the printed word. With the technological boom of the last 20 years we have seen the digitisation of more or less everything that could be digitised, from music to television, and now the book is part of that revolution.

It was ... a small Massachusetts company that finally made the e-book revolution possible, by overcoming the biggest obstacle to a successful device – the screen.

Until now, the book had held out. This is despite text being the first thing ever to be displayed on a computer screen. When personal computers arrived on the scene in the late 1980s, there was talk of delivery of books digitally to your computer screen – an idea later pushed in the early 2000s by American author Stephen King, who charged users to read a novel, *The Plant*, that he serialised online. But the idea was unpopular – users were confined to their computers in order to read the content and the bright computer monitor proved unpopular for reading for any great length of time. King later abandoned the project.

Around this time, the first e-book readers appeared on the scene. But these early devices, the RocketBook and Softbook, had flaws that meant they could never be considered a serious challenge to the printed word. The bulky and heavy devices were still fitted with backlit LCD screens for reading, causing the same problems with eyestrain that computers have – and meaning that the battery life clocked in at a paltry two hours or so. The readers still had to be connected to a PC to receive content, and the selection of books was poor.

It was, in fact, a small Massachusetts company that finally made the e-book revolution possible, by overcoming the biggest obstacle to a successful device

– the screen. E Ink Corporation, founded in 1997, pioneered a completely new way of displaying text; unlike an LCD, the screens do not have pixels, but rather are made up of millions of 'microcapsules' filled with tiny, positively charged black pigments and negatively charged white pigments, floating in a translucent gel. Different amounts of each float to the top, depending on the charge applied from the electrodes between the capsules, displaying black, white and varying shades of grey.

This was a markedly better way of displaying text. Static pages require no electricity to display them, vastly improving the battery life of devices (they only use energy when turning pages), and, more importantly, the screen reflects rather than emits light; creating an experience much closer to reading an actual book.

Kindle wasn't the first to use E Ink's electronic paper display – an e-reader using electronic paper was released in Japan as early as 2004, when Sony, together with Philips, launched its first reader device, Librié. The Sony Reader followed in 2006. Yet the first-generation Kindle, launched in the USA in November 2007, was the first to marry Wi-Fi technology with an e-ink display, to create an e-reader that had no need for a PC. Despite the $400 price tag, devices sold out in five and a half hours.

The first Kindle also coincided with the launch of Direct Publishing, a new business model that allowed independent authors to publish their work direct to the Kindle store, initially keeping 35% of revenues (this was later increased, following criticism, to 70%, if the author met certain conditions).

However, the Kindle did not immediately become a worldwide household name. The first-generation device was not released outside the USA, and rumours that Apple was set to release a tablet created a buzz of anticipation that one multi-functional device would soon be available. Yet, after the much-heralded release of the iPad in 2010, it soon became clear that regular readers were experiencing the same eyestrain from the LCD screen that they experienced when reading on a computer, stimulating further interest in the sunlight-resistant, matt-screen Kindle. That said, Barnes & Noble has launched NOOK Color and Pandigital, its Novel Color eReader, with the LCD-based devices selling well and blurring boundaries between tablets and e-readers.

In February 2009, the second-generation Kindle was released, with another major innovation being the free 3G connectivity of the product. This allowed users to begin downloading books straight out of the box, without the need for a personal Wi-Fi network. In October 2009 an international edition of the Kindle was finally released outside the USA, after Amazon reached agreement with 3G operators in over 100 countries.

The third iteration of the Kindle was released in 2010, with a higher-contrast screen, battery life of up to a month and a lighter, smaller design. The Kindle Fire, a full-colour touchscreen version with WiFi, was announced in September

2011 and had hundreds of thousands of pre-orders, retailing in the US at $199. It is expected by analysts to compete with Apple's iPad as an e-reader and tablet in one device.

Commercial impact

Book publishers, particularly in the USA and UK, have realised that e-readers must be embraced, and more than four in five book publishers now offer their publications in digital formats.

For the moment, the Kindle leads the e-reader sector by some distance, accounting for the largest proportion of the 12.8 million devices shipped in 2010, according to research house International Data Corporation (IDC). This was helped by substantial above-the-line advertising campaigns. In addition to the Kindle, Barnes & Noble, Pandigital, Hanyon and Sony accounted for much of the e-reader market growth, which IDC stated rose 325% from 2009's figure of three million units shipped.

In the fourth quarter of 2010, IMS Research stated that the Kindle had taken a 59% market share. Barnes & Noble, through its NOOK device, accounted for a further 11%, with Sony (5%), BenQ (4%) and Hanvon (4%) among the leading competitors. Furthermore, IMS predicted in April 2011 that sales of e-readers would increase 120% in 2011, to 26.2 million units.

Book publishers, particularly in the USA and UK, have realised that e-readers must be embraced.

The market-leading Kindle.

The Kindle, of course, benefits enormously from being owned by and promoted through the world's largest online retailer, Amazon. com, with a nearly seamless purchasing process for users. Barnes & Noble, the world's largest bookseller, is also well placed. Sony too can rarely be discounted as a technological force for change. And China represents a significant opportunity for Hanvon – the short name for Chinese technology company Hanwang Technology Co. Ltd. – and Taiwanese company BenQ's devices.

Amazon remains coy about sales of the Kindle, but it has been estimated that over

500,000 were sold in the first year, and some anonymous company insiders reported that unit sales rose above four million in 2010.

As for the sales of the e-books themselves, Forrester estimated in November 2010 that the value of sales of e-books in 2010 would hit $966m. The analyst predicted that by 2015 the figure would rise to $3bn. Such figures would mark a major shift in purchasing habits – and even the beginning of the end for the ubiquity of the printed word.

What happened next?

For many customers, their initial reservations about electronic books have clearly been assuaged, thanks to the evolution of e-readers and the continually growing selection of titles. Amazon's Kindle Fire can store either 6,000 books, 10 movies, 80,000 apps or 800 songs on its 80GB hard drive. Users of Kindle e-readers in the UK can select from more than a million book titles. Meanwhile, Barnes & Noble in the USA can boast two million titles available to its e-reader users. Combined with increasingly competitive pricing, improved battery life and the lighter, yet equally durable, range of devices, the rapid propagation of available titles will undoubtedly eat away at any major remaining advantage the book has over its digital cousin.

The e-reader revolution is only just beginning.

The readability and indexability of digital devices could also, potentially, mark the realisation of the long-heralded concept of paperless communication, such as in offices and educational institutions. The ability to annotate pages on screen could allow readers to engage with study books and official documents through the device, potentially reducing the high density of paper matter that currently characterises some sectors, such as law.

However, this revolution hasn't started yet, as the e-reader is predominantly adopted as a leisure device. The Kindle Direct publishing model also threatens to break down the old relationship between publisher and author, granting greater freedom (and higher profits) to writers, while diminishing the necessity for publishers.

When publishing a work in physical format, having a publisher is absolutely necessary for success – they take a large slice of the royalties, but pay for the printing, distribution and publicity essential for the launch of any book. E-books can be infinitely copied and the costs needed to publish them are essentially zero, meaning that independent authors can bypass publishing houses and keep a more substantial proportion of the royalties for themselves.

But we should not underestimate the incredible durability of the traditional book. Worldwide, the paper book still dominates, and has done in more or less the same form since the invention of the printing press in the year 1440. It still has advantages over the Kindle that perhaps can never be addressed – it never runs out of battery life, and it provides a sense of tangible ownership that e-readers will never quite replicate.

It seems unlikely that parents will start reading their children bedtime stories on a Kindle and forgo the pleasure of their offspring turning the pages with them, or that art and design books will be replaced on a coffee table by the display of an e-reader. Furthermore, for many readers, the sharing of much-loved tomes between friends and family members is a key part of the reading experience. What is clear though, is that the e-reader revolution is only just beginning.

Picture Credits

p.5: Getty Images; p.12: Procter & Gamble; p.16: Bausch and Lomb Corporate Archive; p.21: courtesy of NASA/JPL-Caltech; p.22: NASA; p.28: Science Photo Museum; p.30: DoD photo by Gunnery Sgt Michael Q. Retana, US Marine Corps; p.35: ROCC Computers Ltd; pp.39 and 40: images courtesy of the Pittsburgh Brewing Company's archives; p.44: reprinted with permission of Alcatel-Lucent USA Inc; p.49: Getty Images; p.54: Advertising Archives; p.60: courtesy of Xerox Corporation; p.65: courtesy of Waitrose Ltd; p.70: reproduced with permission from Raytheon Company; p.76: Fire Fighting Enterprises Ltd; p.79: photo courtesy of DuPont; p.84: Advertising Archives; p.88: courtesy of MIT Museum; p.95: NESTA; p.105: www.vintagecalculators.com; p.109: photo courtesy of Southwest Media; p.116: MCI; p.121: Omron; p.122: BT; p. 125: Computer Museum; pp.130 and 133: Toyota; p.136, left picture: University of Nottingham; p.136, right picture: courtesy of FONAR Corporation; p.139: Ampex; p.146: IBM; p.153: courtesy of IBM Corporate Archives; p.157: image provided courtesy of the Naval Research Laboratory; p162: photo courtesy of the Chinese University of Hong Kong; p.167: Bausch and Lomb Corporate Archive; p.173: Sony; pp.181 and 182: courtesy of IBM Corporate Archives; p.187: Philips; p.190: photo courtesy of 3M; p.195: Philips; p.196: Sony; p.200: photo courtesy of General Electric; p.205: photo courtesy of Kodak; p.210: Renault; p.214 and 215: Dyson; p.218: HP Communications; p.224: UPS; p.229: Motorola; p.232: Blackberry; p.238: Google; p.244: Google; p.248: Octopus; p.254: courtesy of Caltech Archives; p.255: picture courtesy of Toyota (GB); p.260: Apple; p.266: Google; p.272: © 2010 Amazon.com, Inc or its affiliates.